The Plasticology Project

The chilling reality of our
plastic pollution crisis and
what we can do about it.

Dr Paul Harvey

First published 2022 by Paul Harvey

Produced by Indie Experts P/L, Australasia
indieexperts.com.au

Cover design by Daniela Catucci @ Catucci Design
Edited by Anne-Marie Tripp
Internal design by Indie Experts
Typeset in Mr Eaves by Post Pre-press Group, Brisbane
Photo credits: Cover: Romolo Tavani/istockphoto.com; Internal images: Kris-Mikael Krister/unsplash.com, alenaohneva/depositphotos.com

ISBN 978-0-6455040-0-2 (paperback)
ISBN 978-0-6455040-1-9 (epub)

For Phoebe, for helping to shape my love of the natural world and encouraging me to write this book.

Contents

Foreword   vii

Introduction   1

1 Plastic   15

2 The Global Plastastrophe   27

3 The Links   37

4 Ocean Plastics are a Global Challenge   47

5 Rivers – the Missing Link   61

6 Plastic Animals   77

7 Miniscule Plastic   91

8 Plastic Pollution at Extremes   105

9 Space   117

10 Plastics in Soil   127

11 Plastic Vectors   139

12 Some Plastics We Can Live Without   155

13 The Solution to Pollution   167

14 The Corporate Drive to Reduce Plastic Pollution   179

15 The Plasticology Project   189

   What Next?   205

   Notes   209

   Acknowledgements   223

   For More Information   227

   About Dr Paul Harvey   229

Foreword

Hope for the Future

This book is not all about doom and gloom. It draws together the current knowledge and truths about plastic pollution on a global scale, and highlights the effort that is being made to overcome the challenges we face. In this book, I have brought together a wealth of information from published peer-reviewed academic literature, news reports, media releases, special broadcast series and, most importantly, first-hand experiences of people on the frontline of combating plastic pollution. We must first understand the problem before we can have hope of fixing it and it is the aim of this book to showcase what we know – good news or bad news – so far. I have attempted to capture as much information as possible, however in such a dynamic space, there will always be gaps. Each piece of information presented here is a thread in the rich tapestry that is the global effort to combat plastic pollution. We are fortunate to live in an era that benefits from the hardships and lessons learned by previous generations. We have an opportunity and an ability to make a positive change and create a better world for ourselves and for generations to come. This book is a starting point for that future.

Dr Paul

Introduction

E very civilisation that has ever existed on Earth has left something behind after is has finally gone. Whether it is an artefact, a monument or a colossal building, these tokens of civilisations past open a window into the lives of those times. The ancient Romans left behind an empire of buildings, monuments, artefacts, texts and stories that help us to vividly understand the life and culture of those people. We learn from these intricate details about life in the Roman Empire, and from that we can construct a history over many thousands of years. We can get an intimate understanding about home and community life, and we can learn from these artefacts about manufacturing, engineering, design, and even the textiles of the time.

In a similar way, centuries of Chinese emperors left behind opulent temples and great walls that tell of their power and their drive to expand their empire. We are left with delicately painted pottery that tells us stories of life in the dynasties; we see paintings and carvings that all tell a story of life long-ago lived.

In South America, the Incas of Machu Picchu left behind great monuments high up in the mountain tops that tell a tale of life at the extreme. They tell a story of a civilisation that understood complex building techniques and that had a deep connection to the night skies above. Such great civilisations, such time-defying legacies have been studied for centuries to understand ways of life and how these have impacted on our world today. Reflecting back on these times in history, we are presented with defining symbols of those times: the Romans and their Colosseum, the Chinese and their Great Wall, and the Incas and their citadel. These all serve as reminders of those civilisations and characterise those times.

In our contemporary era we have seen great change, from the discovery of the internal combustion engine to nuclear weapons, and now to the development of powerful computers capable of operating effortlessly in the palm of our hand. In our era, often referred to as the Anthropocene (the time of humans), we have the power to control what is happening on the other side of the planet simply with the push of a button on our smartphone. We can control the lights in our living room at home via our smartphone while sitting in a pressurised airplane cabin high above the surface of the Earth. We can walk into a room and with just our voice activate the lighting, turn on the television and the radio, and close the window shades. We can call up and understand the most intricate details of our world simply by typing a few words into an internet search engine. We can diagnose and treat illnesses unknown to those one generation ago using the power of computers and the technological age.

We are the civilisation that can transport ourselves around the world in less than a day, and experience cultures that less than half a century ago we could have never dreamed of crossing paths with, even in our wildest imagination. Today, we move food from South America to Africa in a matter of days, we have sent ourselves to our nearest celestial neighbour – the moon – and we have even now captured in a single image the most powerful place in our existence – the centre of a black hole.

But from all of this, what will be the hallmark and iconic remnants of our time that will be carried over into the world for the next great civilisation? In centuries to come, what will archaeologists and anthropologists be discussing as the pivotal defining feature of our era? What will *we*, the civilisations of the post-industrial era, leave behind for future generations? Will future generations look on the current era – the Anthropocene – as a time of great advancement with time-defying legacies?

The Environmental Movement

Rachel Carson, in her landmark 1962 book *Silent Spring*, offered the world a window into the future of our planet, our environment and our civilisation.[1] Carson made stark observations about the impacts of pesticide application on the world around her. She poignantly detailed the danger that humans and Earth faced if we continued to treat the planet and the natural environment with the same flippant contempt displayed when indiscriminately applying chemicals to our agricultural and natural areas. Carson faced huge criticism when her work was published, with governments and industries alike scrambling to discredit her observations. Put simply, Carson's observations presented an inconvenient reality for governments and industries around the world who wanted to persist with their new-found products of environmental mass destruction.[2] Why? The double bottom line was profit and convenience. Looking back on Carson's work, many who seek to care for the environment identify that time as being the beginning of the modern-day environmental movement.

The release of *Silent Spring* was timely. It came 10 years after the Great Smog of London in 1952 which killed thousands of Londoners,[3] and shortly before the 1967 *Torrey Canyon* oil spill which dumped up to 36 million gallons of crude oil off the coast of the United Kingdom and France, leaving behind an oily slick still considered one of the worst oil disasters in the history of the United Kingdom.[4] This period left global leaders with little choice but to begin seriously considering the environment in legislation and protective instruments. The following decade saw an unprecedented number of countries centring the environment in their national political agenda. From the establishment of the Environmental Protection Authority in the United States of America to the Basic Law for

Environmental Pollution Control in Japan, there was a global political push to undo some of the damage already caused to the global environment.

Despite the progress made in the 1960s and 1970s to instil in the global conscious a sense of importance regarding the preservation of the natural world, humans are very slow learners and are often quick to forget. As a result, we now face a new generation of environmental problems previously unseen by any other civilisation before. These problems include loss of habitat and mass extinctions on all continents, irreparable damage to human health as a result of chemical emissions into the environment, a global climate changing at a rate unprecedented in history, and a daily struggle in many parts of the world to find sufficient nutritious food and clean water. This is all a result of over allocated, insufficient, or poorly managed resources.

Poor management of resources has played a huge role in shaping the global environment throughout the 20th and 21st centuries. With the exception of a small few, the biggest and most catastrophic environmental challenges that the planet faces today are a direct result of the use (and misuse) of resources and improper management of the waste stream generated from those spent resources. Climate change is a direct result of poorly managed carbon resources; soil pollution in New York City is a direct result of poorly managed industrial resources; bleaching of the Great Barrier Reef off the coast of Australia is the direct result of poorly managed agricultural and industrial resources as well as climate change; and the depletion of fish stock in the Mediterranean Sea is the direct result of poorly managed marine resources. The list goes on and is seemingly endless. In recent years we have become much more in tune with another global environmental challenge born from the use and misuse of resources: plastic pollution.

Plastic pollution was first documented as a potential problem in 1971. Marine biologist Ed Carpenter was on a research cruise in the Sargasso Sea – a region of the Atlantic Ocean – when he noticed little white flecks floating along in the seaweed. After carefully examining the particles, he realised that they were fragments of plastic. In 1972, in the journal *Science*, Carpenter introduced plastic pollution to the global scientific literature.[5] Carpenter documented 3,500 pieces and 290 grams per square kilometre of plastic in the Sargasso Sea. This represents an astonishing amount of plastic given the relatively young age of consumer-use plastics in 1971 and the crude, by today's standards, analytical capacity of the time. Today, as analytical techniques have improved and we can now see much smaller and finer particles with the aid of microscopy, we could expect to discover an even greater number of plastics in this area. Carpenter's observations and environmental impacts of plastics were captured by his warning about the need to limit plastic production and curb plastic pollution. He wrote:

'Increasing production of plastics, combined with present waste-disposal practices, will undoubtedly lead to increases in the concentration of these particles.'

Just as we heard the warning from Carson about organic pollutants, we heard a very strong warning from Carpenter. But did we heed that warning and go about making a change before the world became flooded with plastic pollution?

In 2017, 45 years after Carpenter made his discovery, the BBC released the Blue Planet II documentary series that examined marine plastic pollution.[6] The series was narrated by Sir David Attenborough and explored some of the natural environments that are most impacted by plastic pollution. In the series, Sir David Attenborough remarked:

'Industrial pollution and the discarding of plastic waste must be tackled for the sake of all life in the ocean. Surely, we have a responsibility to care for our planet. The future of humanity and indeed all life on Earth, now depends on us.'

In recent times we have heard other calls to action from prominent conservationists, including Dr Jane Goodall, Leonardo DiCaprio, and the editors of the National Geographic Magazine through its Planet or Plastic? campaign.[7]

Today, 47 years since Carpenter's original observations, we have up-scaled plastic manufacturing at a greater rate than ever before. Plastic is now all around us, from the food we eat to the cars we drive – plastic is a part of them all. The global population has grown rapidly in the time since Carpenter's work, and so too has plastic consumption. Waste has become an ever-increasing problem, as we do not have a solution to the waste generated once the plastic is no longer of use to us. Plastic is discarded carelessly, haphazardly and irresponsibly all over the world and has created an environmental catastrophe that we are only now, as a civilisation, waking up to. Throughout the world, we see entire communities engulfed by plastic pollution, the loss of wildlife and the destruction of habitats resultant of plastic pollution.

Born into a Plastic World

Rather than all of the great advancements in technology, global connectivity and blending of cultures that the Anthropocene has delivered, perhaps the current era will be remembered for a much darker legacy. Will we be remembered as the plastic pollution generation? I pondered this question recently when holding my hours-old niece in my arms. A precious new life, completely unaware of the enormity of the challenges that she will face in her lifetime. A child born in the year 2020 will have lived a long, full life in the year 2100.

I remember hearing stories told by older relatives

born in the early 1900s about 'the way the world was back then' and about how different the 'modern' world is from when they were a child. I recall in particular the stories from my grandmother (born in 1920, exactly 100 years ago at the time of writing) of living in a 'hut' with dirt floors in a remote town that is now less than two hours' drive from bustling Sydney, Australia. She would reminisce of times gone by that were much simpler and more relaxed. She would tell of childhood days playing by the river or of hours spent exploring the bush with some of her 11 siblings. She would comment that 'we didn't have all of these fancy gadgets back then'. Of course, we take these stories with a pinch of salt because, as we know, time and nostalgia allow us to view history through rose-coloured glasses. A world without 'fancy gadgets' also meant a world without improved sanitation, a world that was in the grip of two world wars and their aftermath, a world where food was scarce and medical science was rudimentary at best.

But the point here is not to compare the reality of the bygone era with nostalgic memories, the point is to show how different the world is today compared to 100 years ago. In less than a century, the notion of living in a hut with a dirt floor in Australia is almost unheard of. The childhood home of my grandmother (if it were still standing) would be accessible by a motor vehicle with an electric solar-charged engine, air-bags, self-guided navigation system, lane departure warnings and heated seats! Imagine what my newborn niece will be telling her family of her upbringing in the century beginning in 2020.

As I sat cradling the precious new addition to the family, I cast over her a hope for her future and the world in which she is to grow up. It is my hope for her that she can grow in a world that learns from mistakes, a world that seeks to be better, and a world that is sustainable for the next 100,000 years. I hope this for her because behind the vibrant veil of technological success and advancement of the latter half of the 20th century and through the 21st century, there is a world that is struggling to survive.

The current trajectory of the world, of planet Earth, is – in one word – disastrous. In the 100 years of rose-coloured nostalgic memories that I heard from my grandmother, not once was there a memory of the way the Earth has changed from an ecosystem in balance and one that can service the needs of its inhabitants, to one that is in peril and struggling to survive. At the current rate of global population growth and environmental destruction, the year 2100 seems a pipedream. We, as a global community, are so selfish, so in the 'now' moment, so obsessed with advancing the human race, that we are stripping the Earth of everything that it and *we* need to survive. We are placing a pressure on the Earth that means that humanity will not be able to continue indefinitely.

It is my belief that very few people ponder what it *actually* means to be sustainable for future generations. The term 'future generations' does not simply mean our grandchildren, it means their grandchildren, and their grandchildren, and their grandchildren and so one after that. Generations of humans (our great, great, great, great grandchildren) will depend on the planet that we are currently inhabiting and calling home. Future generations may well look back on the current era as a time when humans caused irreparable devastation and damage to our planet, changing its entire direction and leading to the end of an entire civilisation. Is it too late to make a difference to reduce the impacts of plastic on planet Earth? Have we, as a civilisation, pushed planet Earth to a breaking point with our consumerism of convenience? Can the damage that *we* have done to the planet, its ecosystems and its biodiversity be reversed? Or has that opportunity passed?

CHAPTER ONE

Plastic

Plastic is an incredibly versatile material. Despite its very humble beginnings and slow market uptake, plastic has become one of the most important manufactured resources that has ever been introduced into the modern world. Try walking through a supermarket without passing a single piece of plastic. It is impossible to do as most products will be packaged, wrapped, or protected by plastic. If the products aren't packaged in plastic, then they are likely sitting on shelves and display cases that contain plastic. If you still can't find plastic there, then take a look at the fit-out of the shop. Wiring, flooring and even lighting will all contain plastic in some form or another. Wire, for instance, is encased in plastic that acts as an insulator, preventing one from touching another, becoming too hot, and being an electrocution hazard.

The world around us is full of plastic, and many of our day-to-day activities could not exist without the aid of plastic. Plastic, whether we like it or not, is an integral component of our existence in the same way that water and oxygen are integral to our lives. Although our ancestors may have lived their lives in the absence of plastic, as we have evolved over the last century, we have become intimately dependant on plastic. Plastic has engulfed the modern world and we, as a civilisation, have become dependent on it as a resource.

A Brief History of Plastic

Plastic is not just plastic. The word plastic originally derives from the Greek words *plastikos* and *plassein* which mean 'to mould'. Early plastics were formulated on exactly that principle, the idea being to have a material that could be moulded and shaped as desired. The origin of modern plastic is rather complex and many talented individuals have been credited for creating what we now know as plastic. While many credit Leo Baekeland, the creator of Bakelite in 1907,[1] as the master of modern plastic, there were actually murmurings of the new technology 60 years earlier in 1855. The earliest records of a plastic-like material having been developed are from 1855–1856, when Alexander Parkes developed a celluloid material that he named Parkesine. This material was used for the production of a range of prototypes and concept pieces, many of which were the precursors to modern day plastic applications. Unfortunately for Parkes, his business was slightly before the times and in 1866 went broke.

Meanwhile, inspiring the work of Baekeland was a series of other chemists and inventors who provided the precursors to modern plastics. One such inventor was the very talented and opportunistic John Wesley Hyatt, who set about answering the call of a New York-based company that manufactured billiard balls. The company, hindered by the rising cost and unstable supply of ivory, wanted a replacement product that was just as durable and attractive. Drawn by the financial incentive, Hyatt approached the challenge with a number of ideas. While progressing with experiments, Hyatt discovered that a mix of nitrocellulose, camphor and alcohol produced a material that could be suitable for a wide range of applications – even if they weren't very suitable for billiard balls. This material, which he called Celluloid, was used for creating hairbrush handles, piano keys and – something now lost to history – detachable shirt cuffs and collars. Fortunately for Hyatt, the substance was very difficult to degrade, making it very stable and a successful material for long-term use.

Inspired by all of the work on very early plastics, Baekeland began trying to optimise the production process and in 1907 applied for a patent on a process,[2] later described in a seminal paper in 1909, for a formaldehyde- and phenol-containing product heated using an oven-like apparatus called a Bakelizer. This product became known as Bakelite and its invention was an era-defining moment of the 1900s. It had a wide range of applications in industrial settings, such as electrical insulators and casings, and in the home for kitchenware, kids' toys and even jewellery. Bakelite plastic experienced a huge surge in popularity among the 'elite society', replacing ivory in things like billiard balls, piano keys and even saxophone mouthpieces. Plastic soon became viable for commercial production by the Bakelite Company and was suddenly an accessible commodity for the wider community. Very quickly, plastic became the latest must-have technology.

The process of making plastic has changed over the decades thanks to the many talented scientists who have been able to develop plastics fit for almost any purpose. But early plastic researchers would still recognise some of the key steps in the process of manufacturing most contemporary plastics. The same basic process is used, as I'll explain. Stay with me here, the chemistry may become a bit confusing. Modern plastic production, in almost all cases, uses hydrocarbon (fossil fuel) derived monomers (a molecule capable of interacting with other molecules) as the base to create polymers (a group of monomers that have bonded and formed a more complex structure). Hydrocarbons are organic compounds consisting of hydrogen and carbon. They can be very simple or highly complex depending on the number of hydrogen and carbon atoms that are present (atoms can be thought of as building blocks). The number of polymers present and their structure within a plastic compound is used to determine its properties. For example, you may have heard of high-density polyethylene, HDPE for short. This is one of the most common types of plastic and is used most widely in bottles, packaging and toys. The structure of HDPE is very simple and is a chain of repeating, joined monomer units of ethylene (C_2H_4). Ethylene is initially a component of natural gas (a hydrocarbon) in the form of ethane, and requires heating and exposure to steam in a process known as steam cracking before it becomes ethylene. Once in a usable form, ethylene undergoes a catalyst step (how the monomers are joined via the process of polymerisation) that can be either chromium/silica, Ziegler-Natta or metallocene – the particulars of these processes are not useful to the story so I won't dive in too deeply. The resultant plastic chemical structure is a long string of joined carbon-hydrogen polymer units that has very high tensile strength.

Modern plastics have been created for all types of applications. The group of plastics that HDPE sits in – polyethylene plastics – has four subcategories: low-density polyethylene, medium-density polyethylene, high-density polyethylene and ultra-high molecular weight polyethylene. Generally, there are seven groups of plastic, including: polyethylene terephthalate (symbol 1), high-density polyethylene (symbol 2), polyvinyl chloride (symbol 3), low-density polyethylene (symbol 4), polypropylene (symbol 5), polystyrene (symbol 6), and others including polycarbonate (symbol 7).

I have listed all of these plastic categories here because I want to set you a challenge. Without any further description of these plastics, see if you can identify them in your daily life. I guarantee that you will be able to find them all – possibly in places that you least suspect. The way that you will be able to tell is the number symbol on the plastic – for example, you may see a number two for HDPE, or a number three for polyvinyl chloride. This is an important point to make about modern plastics because these symbols also provide an insight into how easily the material can be recycled or re-purposed. For example, polyethylene terephthalate (PET) is widely used in soft drink bottles and is relatively easily recycled. In contrast, plastics like polycarbonate can be a much greater challenge to recycle, particularly when they contain the compound bisphenol-A, an endocrine-disrupting compound. Few jurisdictions anywhere in the world will recover these materials – those that fall within the number seven category – for recycling.

So here we have reached the key question about the problem of plastic. As I will explain in the next section, plastic is a manufactured problem. It is the way plastic is made, the way it is used as a resource and the way it is managed after it has completed its function that makes it one of the biggest environmental challenges of our time.

A Manufactured Problem

Given the great versatility of plastic, and the hugely beneficial role that it has played in the realisation of the modern world, what then is the problem with plastic?

Plastic is a diverse material that has an infinite number of applications and often significantly outperforms competing materials. It is lightweight, can be shaped, is often temperature tolerant and is highly durable. But it is this robust performance as a material that leads us to why plastic is a problem – environmentally, anyway.

Environmentally, plastic is a disaster. As we will explore in the book, plastic has found its way across the globe and into every single nook and cranny possible. But in the simplest terms, it is the way plastic is made and the way that this resource is managed throughout its life cycle that makes plastic so difficult to deal with. As discussed in the previous section, plastic is made from a huge range of different chemicals and compounds. Over the last century, plastics have become much more chemically complex as scientists strived to create materials that perform better and for longer than their predecessor.

One of the biggest problems we face with plastic stems from the complexity of the material and the way it is manufactured, and our very limited knowledge about what to do with it once its useful life has expired – do we recycle, reuse or dispose? Linked to this is that we have very little knowledge about the time it takes for plastic to break down once it is no longer used and is disposed of.

Disposal of plastic – whether that be formally in a landfill or informally in a community dump site or through the reckless act of littering – leads to the one and same challenging question: how long will the plastic stay in the environment? We know just by looking at our local creek or river that plastic hangs around for a very long time.

Whether it's a plastic bottle, food package or kid's toy, you can probably see them all on the banks of the creek, and the only place they will be going is downstream into the ocean. Although we know that the plastic will eventually become brittle and probably break up into smaller pieces, what happens after that is much less understood. This leads to the point – and the problem – that we don't really have a good grasp of how plastic breaks down in the environment, nor how long it takes.

Some researchers claim that plastic can take anywhere between 10 and 1,000 years to break down.[3] What we actually define as *break down* is also hotly contested. Does break down mean that the plastic is completely non-existent with no trace left behind? Or does break down mean that the plastic has simply altered its physical form such that it has turned into smaller and smaller pieces? Some also refer to the half-life of plastic, the half-life in the biological sense that is. It has been postulated that plastic can break down – take on a completely different chemical structure – in the presence of certain bacteria and organisms. The half-life in this sense refers to the amount of time it takes for half of the plastic to break down in response to these biological processes.

You might be wondering why this is such an important question and why it forms the basis of the plastic pollution problem. The reason is that regardless of which way you cut the plastic pie, the answer still remains the same. Plastic remains in the same form in which it was manufactured, albeit in slightly smaller pieces, for many years after its working life has expired. This is what causes so much contention in regard to the use of plastics in everyday life. This is the basis of the plastic pollution crisis that is currently gripping the world. Once made, plastic is *really* hard to make disappear.

As much of the plastic that reaches an end-of-life phase will end up in the environment, where it will stay for many hundreds or even thousands of years before finally breaking down, this creates an enormous environmental burden that scientists are only really now beginning to fully comprehend. So we are faced with the challenge of how we manage plastic as a resource. How can we use plastic in a way that means we can continue to use it where needed, but in a way that doesn't result in plastic also polluting the environment, without destroying the soils, rivers and oceans of which we have a deep human connection? How can we curb the plastic tide and prevent plastic from being the legacy of this generation? This is the Plasticology Project.

CHAPTER TWO

The Global Plastastrophe

Are we really facing a global plastic catastrophe, a plastastrophe? As I sit pondering this question, I look over at my cup of peppermint tea. I must admit I didn't purchase this tea; it was a 'gift' from a hotel room that I recently stayed in. The packet that the teabag came out of was, you guessed it, made from plastic. I normally don't buy tea that is packaged in this way, so this is a rare moment for me. Looking into my cup, I notice that the teabag seems to be made from some kind of nylon mesh material. That's right, the bag itself is made not from a plant-based fibre, but from plastic. Gone, it would seem, are the days of a teabag being made from natural material that can be composted along with its contents, so that it can eventually break down and return to the soil, ready to become the fertile ground from which new tea will grow.

Is it a matter of convenience that we now use plastic mesh for tea bags? A convenience for whom? Surely the manufacturer who so proudly emblazoned 'certified organic' and 'Fairtrade' across the packaging would be mindful enough of the current global push to curb plastic pollution to stay well away from plastic, favouring instead other biodegradable packaging? While trying to come up with a good answer for why teabags contain plastic, I turned to the ever-trustworthy World Wide Web. Apparently, using plastic makes it easier to close the bag during manufacturing. The plastic melts, allowing for a nice crimped edge to firmly secure the tea within the bag. It makes sense when you think about it: we wouldn't want the tea to be poorly encased in its bag, and we certainly wouldn't want tea to escape out all over our cup as a result of a poorly-sealed teabag.

On the packet, alongside the self-promoting statements about the tea being the finest in the world and the various claims about the importance of drinking organic products that are free from chemicals – which is somewhat ironic given that the plastic teabag will introduce chemicals into the brew – there are instructions on how to brew the tea. We are told by the packet that we must let the tea steep for five to eight minutes. We are then to squeeze the teabag to remove any trace of flavour and then discard it.

So that teabag has an operational life of five to eight minutes. That teabag will then take thousands of years to breakdown. This is, in one teabag, the crux of the plastic pollution problem. This is the problem of a versatile and highly functional product that has a very long degradation period and no adequate solution to its waste stream.

A 2019 study from researchers at the University of Plymouth International Marine Litter Research Unit set out to illustrate this point. The researchers reported on plastic shopping bags that had been purchased and buried in soil a number of years ago to assess their functionality once exhumed some years later.[1] In the accompanying media coverage, shoppers were shown carrying and using these plastic bags as though they were brand new from the store, albeit a little bit dirty. The bags had undergone very little degradation and still performed the role of a plastic shopping bag. The twist, though, was that these plastic shopping bags were not the usual lightweight variety, the shopping bags were made from a biodegradable material that was supposed to disappear over time. Unfortunately, these biodegradable plastics remained in the environment.

This too is another example of the impetus of the current plastic pollution crisis. Despite having a range of biodegradable plastic products to help combat the issue of waste from plastic pollution, they too have not been fully thought through and tested in their waste phases, so here we are again faced with another aspect to the plastic pollution problem. We have a market flooding with products that, on face value, appear to be a solution, but are in fact contributing to the problem.

Plastic pollution may not feel as though it is something that you need to worry about. You may live in a home where plastic pollution is the least of your interests because you have a municipal recycling system that takes your plastic, cardboards and glass away for recycling. All of the other waste products are placed in another bin and sent off for landfill. You may feel that by placing those items that have been traditionally marketed as recyclable into the recycling bin that you have acted in the best way for the environment and the natural world. But do you know what happens when your recycling is collected from the curb in a garbage truck and is taken away?

Where actually is *away*?

International Trade of Plastic Waste

In a 2015 study, China was identified as the largest contributor to global plastic pollution into the oceans, with somewhere between 1.35 and 5.53 million tonnes of plastic being released into the surrounding oceans by the population living within 50 kilometres of the coastline.[2] What the study did not account for was the proportion of that plastic that had been *taken away* through recycling and municipal waste programs from other parts of the world.

For many years China has been the dumping ground for products and materials that the rest of the world no longer wanted. With the hope of being able to turn these materials into recycled commodities, a wave of material would be sent to China on a daily basis. This *was* a solution to the global plastic problem – or so it seemed. The problem with this solution is that it wasn't sustainable. Globally, we have been extremely negligent and short-sighted with the way that we treat the waste stream and end-of-life products that we generate. The reality of this negligence is that a select few countries have become overburdened by the problem of the rest of the world. Ultimately, we can't keep passing off our end-of-life packaging and products to whichever country is willing to accept it. This was realised by Chinese authorities in 2017 when they introduced the Prohibition of Foreign Garbage Imports: The Reform Plan on Solid Waste Import Management. The purpose of this ban was to stem the inflow of solid waste – including plastic – from countries that had become reliant on outsourced waste processing. Prior to the ban, China imported approximately 8.88 million tonnes annually, with much of this material (70.6%) being buried.[3] The impact of the ban on global plastic waste management was significant as the market had to make a drastic shift. The flow of plastic and other solid waste was diverted to other locations around the world. Southeast Asia was the largest recipient region for the waste that was previously destined for China, thus creating an environmental burden in that region. Sadly, without in-country management of plastic and other solid waste and preference over export-management strategies, the issues that were facing China prior to the ban will continue in other localities. There has to be an impetus from individuals, businesses and governments for substantial change to waste management strategies otherwise the problem will continue to be shifted throughout the world.

A Simple Solution to Plastic Pollution

While other solutions to plastic pollution will be discussed, it is worthwhile mentioning here one that was implemented by officials in China to curb plastic pollution. It is not the intention to focus on one country alone in this chapter, but as China has the world's largest population, we can learn a lot from them about the solutions that work and those that need further tweaking.

Having travelled extensively in China, I have often observed plastic shopping bags being used by almost all shops vendors in a very similar way that we see occurring in other parts of the world. It is commonplace at a supermarket for each variety of fruit to be placed in a single-use plastic bag for weighing and labelling. Sure, it is only one plastic bag, what damage can it do? Well, think about the population of China. In 2019, the population was somewhere around 1.42 billion people. If each person went to the supermarket each day and received only one single-use plastic bag, that would equate to 518.3 billion single-use plastic bags used each year in that country alone. What happens to all that plastic and where does it end up? Large amounts will end up in the natural environment, in rivers and creeks. The plastic may also find its way into agricultural systems where it is used as a form of mulch, either to cover crops as they grow or shredded and turned into the existing soil to bulk out the soil material. China is also home to one of largest sources of fluvial transported plastic pollution, the Yangtze River, showing that movement of plastic pollution through the environmental compartments is not simply a hypothesis, but a reality. In response to this growing environmental burden an announcement was made in 2020 that by the end of that year all single-use plastic straws would be banned throughout the country. More critically, a ban on plastic bags will be imposed by the end of 2022.

In a country the size of China, this is remarkable progress and demonstrates what can be achieved when willpower is present. But this is not a problem limited to one country, and while the best efforts of China to combat local challenges will contribute to the overall global reduction in plastic pollution, this is simply a drop in the ocean unless other counties also commit to the same plastic pollution reduction targets.

In this book I talk about a number of challenges that we and our planet are facing due to plastic pollution. I explore the enormity of the problem and some of the solutions that are already being implemented. The plastic pollution battle is far from over, and the global community has a very long way to go before the problem can be deemed solved. But for now, I will say that plastic pollution can only reduce if all counties strive with the same willingness to make a change. We can all learn from countries that are doing well, and those that haven't yet started the plastic revolution. Ultimately, we – both as individuals and collec- tively – can learn from countries like China. As the country with the world's largest population, China is currently a leader and biggest contributor to the global reduction in plastic pollution, simply by hearing the call to action of the global community to end the overuse of single-use plastics and to educate current and future generations about the importance of reducing plastic pollution in our daily lives. Only by reducing the plastic waste emissions from the largest populations on Earth can we ever hope to resolve the global plastastrophe.

CHAPTER THREE

The Links

Throughout history, the ocean has represented the gateway to far off lands. Oceans provided many expanding cultures a means of transport around the world that was relatively fast and easy compared to previous overland methods. Spanish and Portuguese trading routes were predominantly ocean-based, with vessels departing these territories to venture into unexplored lands to trade and build empires. Spanish and Portuguese migration into South America was a result of the ocean trading routes used by the two countries. Their exploration of trade winds and ocean currents in an effort to expedite trade movements led to the realisation and subsequent occupation of those areas. Without knowing it, early explorers were harnessing the power of the global oceans and a very special mechanism that operates within them. This mechanism is the ocean currents and system of circulation.

Climate Change: A Part of the Story

Like all other environments, humans have made their mark on the global oceans. A slightly less casual observation of the oceans will lead you to see that global warming and climate change is having a major impact on the global oceans. In 2010, it was found that global ocean temperatures had risen by 0.6°C over the previous 100 years.[1] Politicians, government officials, business people and those who have a vested interest in disputing climate change may do a great job at casting doubt over the reality of climate change, but there is clear, irrefutable scientific data on the subject. Global climates are in a very delicate balance and this has an impact on the operation of many other environmental systems. Global oceans depend on a similarly precise and delicately balanced set of parameters to operate. Changes in global climates can have detrimental and irreparable impacts. If the global climate becomes too hot and ice in the polar regions begin to melt, thermal currents become disrupted, rainfall patterns change, and land vegetation structures are forced to adapt. If temperatures become too cold, there is a risk of a breakdown of the ocean circulation currents that power tropical weather patterns and ultimately provide the conditions required for the stability of coral reefs and rainforest ecosystems. These hugely biodiverse ecosystems are the lifeblood of our planet, of our existence.

Many of the world's low-lying coastal areas are at great risk of becoming inundated by rising oceans, threatening homes, crops and lives. I recall a holiday that I took with my family in 2012 to one of the Pacific Ocean's most beautiful island nations, Vanuatu. After a day-long car ride and an hour's boat ride, we arrived at an island community which welcomed us with open arms. After the obligatory welcoming ceremony that involved a great deal of singing and a traditional greeting by the island Chief, we were encouraged to explore the underwater world of the pristine coral reef. All the colours of the rainbow could be seen, with fish of all shapes and sizes, along with sea turtles, sharks – everything. I remember thinking at the time how devastating it would be if this precious ecosystem was ever lost. A few years later, in 2015, Vanuatu and the island archipelago suffered one of the biggest cyclones on record. Cyclone Pam hit the capital Port Vila and decimated many of the surrounding islands, displacing 188,000 people.

Although the death toll was low (11 recorded deaths), the cyclone has been described as one of the largest and strongest on record for the region.[2] Cyclones are a naturally occurring part of life in the tropics, but their increased frequency and intensity as a result of a changing global climate has become a very real problem throughout the world. It is understood that climate change is responsible for changing global sea surface temperatures, rainfall patterns and ocean currents.[3] This in turn is driving localised weather[4] instabilities resulting in more frequent and intense storm events. As we will soon discover, these very same global variations that are resulting in increased storm activity are also at play in the battle against plastic pollution.

Plastic Pollution, Climate and the Ocean

It has now become impossible to visit a beach or shore-front anywhere in the world without seeing the impact of global plastic pollution. Plastic is everywhere. From the wave break to the deepest depths, the ocean acts as a sink for plastic. Plastic is transported by the ocean around the world and plastic is degraded in the ocean. The ocean and plastic have an almost parasitic relationship: plastic pollution depends on the oceans for many processes, and yet never returns any positive benefits to the ocean.

While visiting a small coastal town in Wales, United Kingdom, in 2018, I was fortunate to stay in an apartment by the waterfront. The Welsh coastline is a rugged beast, with waves and huge tides constantly pounding the cliffs and coastal embayments. On the odd occasion when the waves aren't pounding the coastline, the sand and pebble beaches are a great place to explore. Aberystwyth faces west towards the Irish Sea, the passage of water that serves to connect the North Atlantic Ocean from the south near Cork, Ireland, to the north off the coast of Glasgow, Scotland. This water passage is a major conduit for seafaring vessels, passenger vessels, cargo ships, marine fauna and even ocean debris. Without fail, each time visiting the beach, entwined in the seaweed on the high-water mark, was an array of plastic pollution. From ropes and nets (called ghost nets – the nets lost or discarded at sea) used by fishing vessels to plastic bottles, plastic could be found everywhere. On each trip to the beach, it was possible to collect more plastic than could be carried in less than five minutes.

Sadly, much of the plastic pollution washed up on the beaches is not derived from Aberystwyth or even Wales, instead it is carried from continental Europe and further on ocean currents. Despite the effort of the environmentally-conscious local community to implement a range of bans prohibiting plastic bags, packaging and straws, this often has little visible impact on the beach-front due to the distal source of the pollution. Much of the litter is also not contemporary. On one day when I was visiting a south-westerly-facing embayment, a piece of litter caught my eye as it bobbed around in the waves. This plastic yoghurt 'drink' bottle package had French text, and as I later discovered was sold during the 1980s. The product had long since been unavailable and had presumably been bobbing about in the ocean ever since. So how does all that plastic end up on the beaches?

In the same way that global oceans have for centuries acted as passages for warfare, trade and transportation, they are now acting as conveyors of plastic pollution. Global sea circulation currents mean that a piece of plastic dropped into the ocean in Australia could, within a few months, end up washing onto the shore in coastal Wales. This is all to do with the thermohaline circulation of the oceans. A very simple explanation of the thermohaline system is that in different parts of the global ocean, network waters are either warmer or colder at the surface relative to waters at depth. Add to this the differences in salinity of the waters and a remarkable phenomenon of water mixing and circulation begins to occur. Thermocline circulation is often described as the conveyor belt of the oceans as it links different surface and deep-water currents through the global oceans. Importantly, the thermohaline circulation is a delicate balance of a range of confounding factors, and subsequently is impacted by the environmental nuisances resulting from climate change as water temperatures, rainfall and ambient temperatures all change – now we can see the connection between climate change and plastic pollution starting to take shape.

The North Atlantic Ocean is a particularly interesting part of the global ocean and thermohaline circulation. Research has shown that water at the high latitudes of the North Atlantic interacts with the interconnecting water in the Norwegian and Barents Seas. The behaviour of the currents and thermohaline circulation in these areas tends to act like a suction vortex drawing floating debris further north into the Arctic waters.[5] No doubt this phenomenon is instrumental in much of the transport of plastic debris as it moves through the Irish Sea from northern continental Europe, north up the coast of Wales and often deposited on the Welsh coastline. In this way, the global oceans are effectively a plastic pollution highway. A piece of plastic packaging, a drink bottle, or a fishing net can be caught in a global ocean current and be transported from one ocean to the next around continents and eventually the world.

CHAPTER FOUR

Ocean Plastics are a Global Challenge

Plastic pollution, Da Nang, Vietnam.

The North Atlantic Ocean is not the only place where plastic pollution is problematic. On a trip to Da Nang, Vietnam, in early 2018 for a workshop on marine pollution response, it struck me how there was no discussion on the topic of ocean plastics. One afternoon when I had spare time, I decided to walk along the foreshore of the river that transects the city. Along the footpath, teetering on the edge of the water, there were plastic straws everywhere – hundreds of plastic straws. There was also fast food packaging, foam boxes and fishing debris. Walking along the foreshore, it became apparent that plastic pollution was an issue of scale in Da Nang, seemingly exacerbated by the tourism industry and its desire for single-use plastic bottles and straws.

Plastic pollution in the oceans is truly a global problem. Far North Queensland (FNQ) is often seen as the jewel in Australia's crown. A region with outstanding natural beauty, the World Heritage-listed Great Barrier Reef sits juxtaposed by steep cliffs and beaches bordering the lush Daintree Rainforest – the last remnant of the Australia's coastal tropical rainforests. A region of two seasons, wet and dry, the wet tropics of FNQ really are a unique asset to Australia. For many years this was primarily an agricultural region, with tobacco and sugarcane the dominant crops grown in the area. Tourism only recently became a booming industry in FNQ, and with this came an influx of the local and tourist population. As has been the theme in this book so far, wherever humans are present, plastic pollution soon follows.

One day in the 2019 wet season, when walking along a beach in the coastal township of Port Douglas, I became starkly aware of the condition of that beach. It was littered with plastic debris. To give the beach some credit, my visit was shortly after a cyclone had passed over the coast, creating widespread flooding in townships further south, but this was the tropical north where cyclones come and go like the sun and moon – they aren't a particularly special phenomenon and the townships have been built with a level of resilience to these storm events. This beach was covered with a myriad of rope, nets, plastic bottles, plastic bags and plastic packages. Among the pollution, the iridescent blue of synthetic ropes used for commercial net fishing stood out among the washed-up palm leaves and ripe coconuts. While much of this pollution had been washed up on the coastline in the recent storms, the beach scene spoke of a far greater problem: the amount of pollution that resides in the global oceans, moving around on the currents and waiting for an opportunity to be washed back on shore.

Far North Queensland isn't a highly populated region; most of the population is seasonal, with tourists flocking to the region during the dry season to visit the Daintree Rainforest and Great Barrier Reef. In many of the towns, large high-end tourist resorts line the main roads and the waterfront. Tour boats float in the marinas, poised for their daily journey out to the wonderland of the Great Barrier Reef. Parked along the roads are tour buses, all waiting to take tourists to the tablelands and rainforest. With a seasonal influx of tourists, there is a seasonal influx of waste and pollution.

Australian drinking water is some of the cleanest in the world, and yet tourist resorts still provide bottles of water to guests, and visiting tourists purchase bottles of water from supermarkets and convenience stores. This unnecessary consumption of plastic resources – in the form of the bottle – is exactly the reason why plastic has become a global pollution burden. When presented with a plastic bottle of water, the brain is triggered into thinking that there must be something wrong with the freely available water from the tap. Once that thought is triggered, we are more likely to seek out more bottled water.

So, what happens when the water is consumed? The bottle will be thrown away – possibly recycled or reused, but most likely it will end up in the general waste or even, disappointingly, become litter. When waste is discarded indiscriminately, it will start a life of constant movement and transmission around the environment. Whether that is a result of winds transporting lighter materials over great distances, or water carrying the waste through drainage channels and into the rivers and oceans, this waste will eventually become a part of the environment and will succumb to the Earth's natural processes, being transported according to the will of ocean currents.

Sadly, the beaches of FNQ are not the only beaches in Australia that are covered in plastic debris and pollution. Plastic pollution has engulfed Australia's coastline in a way that is unprecedented. Unlike other environmental pollutants such as oil that are very acute and localised in their impacts – usually only reaching a few kilometres from their point of release – plastic pollution can travel far and wide.

The Commonwealth Scientific and Industrial Research Organisation (CSIRO) released a report in 2016 that showed 75% of rubbish pollution (that is any kind of waste debris) that settled on the Australian continental shoreline at the 175 study sites was made up of plastic.[1] Aside from the massive volumes of rubbish and debris that the study identified on the Australian coastline, it documented that the largest deposits of litter were found near to urban areas. This does not sound all that surprising, given that humans are the exclusive source of plastic emissions on the planet, but what is interesting is the findings of subsequent research conducted on beaches in the remote Arnhem Land region of the Northern Territory that showed masses of plastic pollution on remote and isolated beaches.

The CSIRO study correctly pinpointed humans as the source of the plastic pollution – how else would the plastic get there? Well, as one of the contributors to the

Arnhem Land plastic pollution clean-up project, Dr Jennifer Lavers, commented to *The Guardian*: 'It is likely this waste came from Southeast Asia, but we know at the same time Australia's waste is going over to somewhere else.'[2]

Now let's pause for thought. The clean-up crew, led by marine conservation organisation Sea Shephard, collected seven tonnes of marine plastic from the beach. The beach is over two and a half hours' drive from the nearest township and is in one of the most remote corners of Australia, yet it is still touched by the human hand by way of plastic (and undoubtably other forms of pollution). All of this plastic has rafted onto the beach from (most likely) Southeast Asia. In exchange, Australia gifts Asia with some of its own plastic pollution. This is a phenomenal journey for plastic to take in the oceans and highlights why ocean garbage is a huge global challenge. If you're curious to know exactly how big this challenge is, let's talk about one of the biggest ocean plastic pollution sites in the world: the Great Pacific Garbage Patch.

The Great Pacific Garbage Patch

The Great Pacific Garbage Patch is an impressive name. Think about it, in the Pacific Ocean there is a huge man-made feature that is so enormous that it commands the epithet *great*. Historically, this attribute commands respect: Alexander the Great, the Great Wall of China, the great white shark. In no way could the human endeavour to overcome all environments on Earth be better achieved than by installing a structure into one of the wildest, most rugged oceans. Without really trying, humans have achieved exactly that. The ability of oceans to act as transport superhighways has resulted in the accumulation of the world's garbage in the Pacific Ocean.

The Great Pacific Garbage Patch is actually made up of two areas. The Eastern Garbage Patch is smack bang in the subtropical waters between California and Hawaii, off the west coast of USA. The Western Garbage Patch is located off the coast of Japan. Between the two patches is the North Pacific Subtropical Convergence Zone, where warm water from the South Pacific joins cool water from the Arctic. Together the North Pacific current, the North Equatorial current, the California current, and the Kuroshio current create the convergence zone. Like a big vortex, the sea surface currents and circulations drag the garbage into what is now a 1.6 million square kilometre plume of rubbish leisurely floating around.[3] It is estimated that the Great Pacific Garbage Patch is made up of at least 79 thousand tonnes of plastic debris and around 1.8 trillion plastic pieces, comprising fishing nets and various other plastic debris.[4]

The North Pacific Subtropical Convergence Zone is actually one of the most important ocean ecosystems on Earth. Swimming within the garbage are fish, whales and other marine life. The zone is also vital in providing food for marine life and even plays a role in nitrogen and other nutrient cycling through the growth of phytoplankton and other organisms.[5] The garbage is often so thick that there is very little space left for any marine life. The mix of debris, reduced sunlight and toxins means that oxygen content of the water is often reduced, making it difficult for marine organisms to survive in the region.

I am reminded of the documentary film *Journey to Planet Earth,* which I watched during my early under-graduate studies.[6] The documentary was narrated by a very young, very optimistic Matt Damon. Long before the global plight of plastic pollution had become a topic of household discussion, this documentary featured an island in the Maldives that I only remember as Rubbish Island. Rubbish Island, or Thilafushi as I later found out is its official name, is an artificially constructed island on a former coral reef that is made entirely of the rubbish generated from the popular tourist hotels of the neigh-bouring islands. Boats would deliver the garbage to what is essentially a rubbish dump in the ocean, where it was often burned or simply left to leach its toxic cocktail of pollutants into the surrounding water and reef. Fortunately for the ocean and reef around the Maldives, Thilafushi has become somewhat of a plastic pollution campaign poster centrepiece. Thanks to media coverage and tighter regulations, there is now a much more controlled and organised arrangement with recyclable materials being recovered and reused.

The Great Pacific Garbage Patch is the largest of many other similar areas of floating debris distributed around the global oceans. Managing these enormous flotillas of waste will be an ongoing challenge for decades to come. Despite best efforts to clean some of the debris from the ocean, there still remains a bountiful inflow of plastic. Until that tide of plastic is stemmed, ocean plastics will continue to pose a problem and degrade the marine environment.

Plastic in the Deep Ocean

Thanks to satellites and remote sensing we now know a lot about the surface of the oceans, but we still know very little about the dark depths of the oceans – one of the last frontiers of the human endeavour. Space is another of these frontiers that we will visit later.

You have probably heard people say something like, 'we know more about the surface of the moon than the depths of our oceans.' We have certainly learned a lot about the surface of the moon from various space missions, telescopes and studies of our universe. The depths of the Earth's oceans, however, are notoriously challenging places to understand. They are difficult to access, they are a very harsh environment that supports only the most robust and well-adapted life, and overall, they don't really conjure up the same sense of childhood adventure and awe as a spacecraft travelling to the moon. However, some scientists find the depths of the oceans to be fascinating places.

The Mariana Trench is the deepest point of the Earth's oceans known to humans; at almost 11,000 metres, it is deeper below the surface of the ocean than the highest peak of Mount Everest is above. Over the last five years, scientists have explored the depths of the Mariana Trench using submarine and submersible camera technology. The things that they discovered at the darkest depths of the oceans once again showed just how pervasive global plastic pollution is. In this hostile environment, where only very specialised organisms can survive, the scientists captured images of plastic bags and other plastic debris dating back almost 30 years. While 30 years ago, in 1990, humans were strategically deploying the NASA Hubble Space Telescope, one of the great achievements and advancements in understanding the depths of our universe and one of the final frontiers of human endeavour, humans were also deploying plastic pollution into one of the other final frontiers – the ocean.

The Mariana Trench, the deepest point in the ocean, has been said to be so polluted with plastic debris that it exceeds the rivers in some of the world's most polluted cities.[7] The very same attribute of the oceans harnessed by humans as a means of transport for thousands of years has once again been harnessed to move pollution through one of the Earth's final frontiers.

For many years, we have assumed that the oceans work in a very simplistic way: all pollution ends up at the deepest point. It makes sense: when you drop a spoon in the kitchen sink, it sinks to the deepest point at the plug hole. Unfortunately, you also find all the excess food gunk that you did not bother to clear off before tossing the plates into the water. We, as humans, have a silly idea that if our waste is under the water then it has magically disappeared and no longer exists. If you have ever visited a beach in Sydney, Australia, after a large storm event you will know that just isn't true. The series of deep-water outfalls off the coast of Sydney that are responsible for carrying sewage offshore to 'disappear' are often not as effective as planned, resulting in flotsam returning to the beach, and if you are unlucky enough, your line of vision will be obscured as you dive under a wave.

Until very recently, the working hypothesis has been that all plastic pollution (and other matter) eventually ends up at the deepest possible point in the ocean. Not always the actual deepest point, the Mariana Trench, but the deepest locality within reasonable proximity to the point of introduction to the ocean. We know from the previous discussion that some of the plastic pollution does end up on the ocean floor at the deepest point, but research is beginning to show that this is not always the case. Indeed, this new research shows the separation of plastic pollution based on the density of the plastic material. We see that certain types of plastic sink to depth, whereas some types of plastic will float. Researchers conducted a meta-analysis study that showed, generally speaking, lower-density plastics such as polypropylene and polyethylene are most likely to float, whereas those that are of higher density sink to depth.[8] This knowledge presents a new challenge to understanding the flow of plastic pollution around the global oceans. Suddenly, it becomes inadequate to simplify the oceanic movements and assume that all plastics are transported throughout the global ocean in the same way. We must look deeper into the distribution of plastics in the ocean and what this means for the migration of plastics into marine ecosystems.

CHAPTER FIVE

Rivers – the Missing Link

When I was studying for my PhD, I loved nothing more than finding a way to do something other than the work I was supposed to be doing. This led me to being heavily involved with teaching students. I loved to develop research projects, share my (limited) knowledge with the next generation and visit field sites. One subject that I taught was a final-semester undergraduate subject that allowed students to create their own research project and execute that project themselves. I offered some project ideas and was fortunate enough to have a very eager and intelligent group of students take an interest in what was being offered. At the time, microplastic pollution of the oceans was beginning to gain attention. I suggested to the group of keen learners that perhaps we should look at microplastic pollution in rivers around Sydney. I explained that rivers transport sediment to the oceans, so they must also transport microplastics. The logic was simple and seemed to make sense as a hypothesis, so we set out to develop a project examining microplastics in headwater streams around the Sydney basin.

The role of headwater catchments and rivers as sources and transport mechanisms of microplastic really wasn't considered in the international literature at that time, in 2014, so we had a ground-breaking study in the making. We looked at rivers and creeks in the iconic Blue Mountains region of New South Wales, Australia. Many of these rivers flow into the Sydney drinking water catchment. We sampled bed sediments across 11 sites in an effort to find the microplastics that had the greatest likelihood of migrating downstream. Unlike the microplastics discussed in the microplastics chapter of this book (Chapter 7), in this instance we were targeting plastics around one millimetre in size – much larger than what we now generally deem to be truly *micro*.

Our investigation showed that there was a huge amount of microplastic pollution in those rivers and creeks. On average, we counted 639 microplastic fibres per kilogram of sediment. At the time we had no benchmark for what these values meant, but with the benefit of a few more years of research from around the world, we now know that these values are consistent with other rivers, like the Amazon River.[1] These rivers and creeks were definitely acting as a source and transport mechanism for microplastics and our research clearly showed that.

The results were really the first of their kind and so we agreed to write the results for peer-reviewed publication. Unfortunately, at the time, the lack of enthusiasm from the scientific community around rivers being an important part of the microplastic story meant that the work was never formally published. In 2021, a review was published in the *Science of the Total Environment* that considered 38 published works examining microplastics in rivers around the world. Among the conclusions of the authors was that the study of (micro)plastic pollution in freshwater ecosystems remains in its infancy, and so requires much more attention and work.[2]

I am reminded here of a conversation that I had with James Wakibia, a social activist, environmentalist and plastic campaigner based in Kenya, shortly after I started to write this book. James really understands rivers and the role they play in transporting plastic pollution to the global oceans. James has spent many years documenting via social media plastic pollution in Africa and has campaigned for the bans and cessation of the production of plastic products. While talking with James, he described a familiar beach, one with speckles of blue, red, yellow, green – all colours that should not be in the sand at the beach; all tiny fragments of plastic.

What is most revealing when James talks is his profound understanding of the ecosystem-wide impact of plastic pollution: from the introduction of plastic pollution into the environment from cities and villages, through the great conveyors – the rivers – to the oceans. James made the powerfully insightful observation about the ultimate impact of plastic pollution that finds its way to the oceans: 'If a fish dies in the ocean from plastic, a human being somewhere will feel the effect of that. Either they will consume some plastic from the fish or their livelihood from the fish will go down, so more people will be poor because they cannot live their normal lives from the trading of fish ... everybody will feel the pinch at the end of the day.'

James fully comprehends the impact of plastic pollution on the environment and remarked about the haphazard efforts to remedy plastic pollutions on a global scale: 'You cannot urinate in one corner of the swimming pool and expect the other corner to be clean.'

River Cleanup: A Solution

In contrast to the world of scientific and academic publishing, it would appear that anecdotal knowledge and community observation far exceeds the knowledge generated by research teams collecting and quantifying large volumes of sediment and plastic pollution. In 2017, a group called River Cleanup was formed with the initial challenge of picking up as much rubbish as possible from along a river in 10 minutes.[3] As the challenge gained significant media attention, the concept quickly grew and in 2018, the Rhine Cleanup saw 10,000 people from 60 cities along the Rhine River join together to collect rubbish – much of this being plastic waste. By 2019, the program had grown to 10 rivers across Europe and Asia with a volunteer team of around 40,000 people.

To have this enormous turnout of people taking part in river clean-up activities shows the understanding of communities that rivers are an important source and vector of (plastic) pollutants. It also gives us hope that reducing our plastic impact is achievable. If each of the 40,000 people that were involved in 2019 collected just 100 pieces of plastic, that is 4 million pieces of plastic that will never make it into the oceans! The great news is that we don't even have to be a part of a formal river clean-up program to make a difference, each time we go for a walk, swim or paddle along a river we can collect a handful of plastic and rubbish. This can then be disposed of appropriately in a bin or recycling facility. Each little bit counts and we can all play our part to help out.

Knowledge Is Key

In the grand story of plastic pollution, I still don't believe that enough attention is given to the role that rivers play, but researchers are slowly beginning to understand their importance.

Recently, I led an environmental study tour through the Sydney metropolitan drinking water catchment. The tour began in the Blue Mountains, west of Sydney CBD, and over three days we explored various field sites in the Sydney basin that have played an important role in the story of supplying clean water to Sydney and reducing environmental pollution in the catchment area. One of the sites that we visited was situated behind Australia's iconic

Manly Beach. Manly Lagoon in an intertidal waterway at the northern end of Manly Beach – more precisely, Queenscliff. The lagoon is often cited as one of the most polluted waterways on the eastern coast of Australia. The lagoon looks relatively innocuous to the casual observer. It has trees lining the water edge, and fish can be seen swimming in the clear water of the lagoon. So why it earns the title of being one of the most polluted waterways isn't immediately obvious. But the local council doesn't take the title lightly; all along the edge of the waterway are signs warning of the dangers of coming into contact with the water. The lagoon is in an unfortunate position at the bottom of a local catchment that has a lot of different land uses, including industrial and residential. The lagoon in supplied by a small network of creeks and rivers that act very efficiently to convey anything and everything from the catchment downstream into the lagoon. The various land uses in the catchment have, over time, resulted in the discharge of a huge amount of waste into the lagoon. This includes, among various other things, sewage. Despite not being able to see the pollutants in the water, they are still there in the dissolved fractions, leading to the water being unsuitable for contact.

When I was talking to my tour group at the lagoon, we used a small net to catch some of the surface floating organic matter. While the intention wasn't to catch plastics, I ended up with a net full of small plastic fragments and even plastic beads. Just by looking at the lagoon, I had no

MANLY LAGOON

⚠ POLLUTED WATER

🚫 NO SWIMMING 🚫 NO FISHING 🚫 NO BOATING 🚫 NO WADING

Manly Lagoon and its surrounding floodplain form a natural feature of the local district. The water levels in the lagoon are determined by rainfall, ocean conditions, and releases from nearby Manly Dam, however the water quality is affected by urban pollution.

Poor water quality can make the lagoon unsuitable for human contact.

Manly and Warringah Councils are working to improve the condition of the lagoon and surrounding area. To find out what you can do to help reduce urban pollution, visit www.manly.nsw.gov.au or contact 9976 1500.

Warning sign at Manly Lagoon.

idea that there was so much plastic in the waterway and yet there we were, only a few hundred meters from one of Australia's most iconic tourist beaches, staring down into a scoop-net full of plastic particles and debris. Like the dissolved pollutants, it is the pollutants that you can't always see that can pose the biggest problems. Manly Lagoon is one of those coastal water bodies in Australia that discharges through a pipe directly into the ocean. This type of discharge is common along the Australian coastline, with much of the water entering the ocean unfiltered. The lagoon and pipework essentially behave as a superhighway to discharge pollutants into the ocean. I talk about Manly Lagoon here because this coastal river catchment and lagoon system shows, on a local scale, the way that rivers play a very important role in the transport of pollutants, particularly plastics into the ocean.

I used to teach a course at university about emerging contaminants in our fluvial and marine environments. I would often tell the students that the medications they consume on Monday morning will have flushed through their bodies, down the drain, into the river and out to the ocean by Wednesday. Then by Friday lunchtime they can eat fish for lunch at their local fish and chip store, embellished with their medication from Monday morning. It's a humorous thought experiment that has a serious side to it. Rivers are the arteries of our world. They transport life-giving water but they also transport away the unwanted stuff – pollutants. In the same way that we discharge our medications into the river on Monday and we end up eating them again on Friday, we are also discharging our plastic pollution into the rivers and, sometime in the not-too-distant future, we will be eating that plastic pollution, at a similarly invisible scale. While most people seem to understand the concept of medications moving through the environment, few people seem to comprehend the extension of the thought experiment to plastics. Perhaps people just really don't want to believe that they are eating plastic?

Global River Plastic Inputs – A Modelled Example from China

As was discussed earlier, authorities in China have taken a very strong stance against plastic pollution and its environmental impacts. In recent work, researchers considered the role that rivers play in the mass balance of plastic pollution in the environment.[4] The researchers looked at a range of urban creeks and waterways to near-shore and coastal waters. The researchers mainly focussed on waterways in and around Shanghai and tributaries of the Yangtze River. Surface water samples were collected throughout their study areas, and the plastic particles were extracted from the water. The researchers were looking for the suspended plastics in the water.

What they found was a very clear association between the waterways in the city areas and the plastic particle load in those waters. The creeks most heavily influenced by direct runoff from the city areas contained the highest concentrations of plastic particles. Even more interesting was their finding that microplastic pollution was more problematic in the freshwater creeks than the near-shore and coastal waterways. Essentially, the freshwater areas, the source, contained substantially higher concentrations of microplastics than those other areas downstream. So now we have a question: where is all that plastic pollution from the headwater source waterways going? Is the plastic becoming integrated in the bed sediment of the waterway? Are we, as we use water from the river, diverting the plastic to other parts of the environment or into our food chain? The fluvial compartments into which plastic pollution is being retained is receiving a large amount of international attention.

A study that examined the source of marine plastic pollution in the global oceans showed that rivers played a major role in the source and distribution of these particles.[5] The study found that 90% of the plastic pollution that enters the global oceans is sourced from 10 rivers. Disproportionately, eight of those rivers are located in Asia. The study found that the Yangtze, Indus, Yellow, Hai, Nile, Magna-Brahmaputra-Ganges, Pearl, Amur, Niger and Mekong are the biggest plastic polluters. The Yangtze River, the biggest source at the time of the study, was found to discharge 1.5 million tonnes of plastic into the East China–Yellow Sea every year.

The Yellow Sea, which is the northern region of the East China Sea, is a major fishing territory for much of Asia. With a high catch rate of bottom-dwelling species, pollution of this ocean has the potential to cause substantial problems. Having previously explored the impacts that plastic pollution can have on ocean health we know that the introduction of plastic in such high volumes into the Yellow Sea cannot be good for that ecosystem. Without surprise, studies are beginning to show substantial micro- and macro-plastic pollution of various parts of the Yellow Sea and indicate that this plastic pollution may be having a negative impact on fish stocks in those marine environments. Three studies from 2018 all show that microplastic pollution is a major problem in this region, with one study clearly showing that zooplankton – microscopic organisms that float around in the ocean and act as a food source for higher-order feeders – are ingesting and retaining microplastics.[6]

Knowing that zooplankton are ingesting microplastics is highly disturbing, given that these organisms are the basis of our marine food chain and ultimately our agricultural food chain. Ingestion in these lower-order feeders has major implications for bio-accumulation and bio-amplification of the impacts of ingested plastics in higher-order feeders further up the food chain – like ourselves.

The phenomenon of rivers feeding oceans a steady supply of microplastics is not isolated to Asia. Travelling across the other side of the world takes us to the

Mediterranean Sea. The Mediterranean Sea is a major source of food for an enormous chunk of the European and northern African populations. However, overfishing means that the food supply is fast becoming depleted.

I always find it funny how new information makes itself known to me. One night while watching a BBC documentary, *Mediterranean*, by British author and filmmaker Simon Reeve, a segment caught my attention. Simon was travelling around the coastline of the Mediterranean, and in this episode visited southern Spain's Almería region. Simon met a keen researcher and activist who was clearly very passionate about the plastic pollution that was identified as being derived from the greenhouses in the region that act as the 'food bowl' for much of Europe and the UK. The images that flashed across the screen were of layers of plastic embedded in soil and debris of what appeared to be a dry riverbed. Wind and water erosion had exposed the layers and with a slight brush of the hand the plastic would break loose, creating tiny fragments and particles that would be blown on the wind or washed down the river towards the ocean. Simon was visibly distressed by the quantity of plastic debris that had accumulated in the dry riverbed. This segment made me think a bit more about plastic debris in the Mediterranean and investigate the region a bit further.

Daniel González-Fernández stands out as an authority on plastic pollution in the Mediterranean. Over a number of years Daniel's research has filled a knowledge gap about plastic pollution in the Mediterranean with disturbing predictions and realities. In a 2021 paper published in *Nature Sustainability*, Daniel's research team commented something that is shocking and disheartening in equal measures: 'we have estimated that between 307 and 925 million litter items are released annually from Europe into the ocean',[7] and of that litter, approximately 82% was plastic. This paper – a culmination of years' worth of collecting and counting plastic fragments, modelling dispersal patterns and behaviours and identifying contributing sources of plastic pollution – has computed an estimate that is almost unfathomable.

To help me wrap my head around this number I started to think in terms of populations. In 2020, the United States of America had a population of 330 million. Imagine, at best, a mass of plastic fragments equal in number to the population of the USA entering the Mediterranean each year courtesy of Europe. The real kicker in this story, as pointed out by Daniel's team: 'a major portion of the total litter loading is routed through small-sized drainage basins (<100 km²), indicating the relevance of small rivers, streams and coastal run-off.'

Sadly, the Strait of Gibraltar, the only naturally-occurring passage that opens the almost 4 million cubic kilometres of the Mediterranean Sea into the global ocean, feels the full impact of the plastic input from the surrounding land mass. It is said to have some of the highest levels of bottom-dwelling (benthic) marine litter of any body of water in the world.[8] The gateway to the Mediterranean Sea – the lifeblood of Europe and Northern Africa – has become an underwater garbage dump. In a marine environment already struggling due to over fishing and poor resource allocation, the additional stress that is placed on this ecosystem thanks to plastic pollution will most likely lead to the collapse of this entire ocean ecosystem.

CHAPTER SIX

Plastic Animals

I began thinking about this chapter when I was housesitting for my brother. In the kitchen, there was a plastic fast food container filled to the brim with small plastic animal-shaped figurines. Why would somebody have a collection of small plastic animal-shaped toys in their kitchen? It turns out that they were a giveaway gimmick from a large chain supermarket in Australia, released in response to a competing supermarket that had recently given away grocery figurines as a similar marketing tactic.

The small plastic animals were shaped like the African animal characters that feature in the *Lion King* movie. An irony lost on many would be that the animals that this promotion sought to mimic are the animals that live in some of the most plastic-polluted environments on the terrestrial Earth.[1] These stationary and inanimate figurines that play no functional purpose besides creating brand allegiance for supermarket shoppers are feverishly collected by those who fall for the tricks of the marketing trade. We are told by the talented marketing agents that we need these figurines, so we seek them out.

Over the next few weeks, I began to notice that these figurines even became an economy of their own, with the supermarket releasing a 'rare' variety that had fine nylon (plastic) hairs that acted as the mane for the lion figurine. The figurines were swapped, sold and even auctioned so that those who were collecting them could complete their set. The hair-adorned figurine was reportedly listed for sale on an online auction site for AU$100,000.[2] Ultimately, however, the hype around these plastic animals passed and the world has been left with another plastic burden to manage.

Sadly, less than six months after these figures were released, a small community group dedicated to removing plastic pollution from Sydney's waterways, Splash Without the Trash, posted images online of the figurines that had begun polluting the marine environment. As the world becomes more aware of the impacts of plastic pollution and grapples with the implications, Australian supermarkets are promoting themselves using small plastic figurines. Is this clever marketing or environmental negligence?

The Role of Animals in the Transport of Plastic Pollution

From animals that are made of plastic, to animals that consume and transport plastic through the environment. It is no secret that animals play an integral part in the transport of material around the planet. Look, for example, at the great African elephants, who spend their days grazing, leaving behind deposits of undigested food and fertiliser as they move on their daily migration. Being herbivores, seed and plant material is included in the material that they leave behind. Elephants are often stalked by horn-billed birds who scavenge through the faeces hoping to catch a tasty morsel of the food left behind. What isn't gathered by these birds is then left on the ground to eventually germinate. While the African elephants do this with such ease, they are not alone: almost every animal on Earth acts as a vector for the distribution of seeds and plant material, which results in the spreading of species across the planet. This is one of nature's great mechanisms for self-perpetuation, and it has a role in the plastic pollution story.

Unfortunately, the natural processes that are so beneficial for the dispersal of species are also perfect mechanisms for the distribution of unwanted materials. We are all by now familiar with the images on social media, the evening news, or in the film that the local sustainability movement showed last month, that depict a sea turtle having a plastic straw extracted from its nostril after inhaling it while swimming or attempting to eat it as food. Perhaps if you haven't seen that image, you have seen footage of a person sifting through the intestinal contents of a dead seabird that was unfortunate enough to have consumed a mountain of plastic fragments after mistaking them for food items on the near-shore coastal fringe.

This, however, is a reality. This isn't just Hollywood special effects or post-production where the editor digitally adds in plastic fragments to tell a sensational story. Our global sea animals are at risk.

It is estimated that more than 55% of seabird species are at risk from interactions with plastic pollution.[3] Global seabird habitats are overwhelmed by some of the highest rates of plastic pollution accumulation in history.[4] For many years, research conducted on Lord Howe Island, off the coast of Australia, has demonstrated the devastating impacts of plastic pollution in our global environment. Populations of wedge-tailed shearwater have been monitored and their stomach contents documented. The stomachs of these magnificent birds have constantly been found crammed full of plastic fragments, from bottle caps to plastic shards, that they have ingested. In a long-term study between 2005 and 2018 examining fledgling stomach contents, up to 52% of the population contained plastic fragments.[5]

The wedge-tailed shearwater is an interesting seabird in that it appears to be one of the largest consumers of environmental plastics.[6] The reasons for this aren't yet entirely clear, however it is believed that the feeding behaviour of the birds, and the colour, shape and even the scent of the plastics may all be contributing to their high-plastic diet. One study that considered the stomach contents of birds that were both near-shore and off-shore showed a difference in the types and even colours of the ingested plastics. It was proposed that distance to human populations may be a contributing factor to this distribution, or even a colour preference of bird species in difference environments.[7]

The wedge-tailed shearwater of Lord Howe Island aren't the only seabirds in peril. Seabirds from all over the global oceans, from the Laysan albatross and Bonin petrel of the Midway Atoll to the Newell's shearwater in Hawaii and even the northern fulmar of the Labrador Sea, are becoming impacted by the influx of global marine plastic pollution. The global impact of marine plastic pollution on seabirds has become so concerning that researchers around the world published over 1,230 articles on the topic between January and October 2018.[8]

Seabirds cannot digest plastic, and if they swallow too many of these particles then their digestion will soon become impaired. In short, there won't be any room for actual food in the stomach of the bird. In addition to the physical blockage, studies have shown that the plastic particles in the stomachs of seabirds have the potential to release a range of toxic and harmful chemicals into their bloodstream, potentially causing it further health complications. Should the seabird be lucky enough to pass these plastic particles, or in the unfortunate event of the seabird's death, these plastic particles are then returned back to the food chain where animals that feed on the droppings of birds[9] or dead bird carcasses may consume them. Seabirds are but one of the many examples of plastic pollutants accumulating in animals, and the risks this may pose for species population health.

Sea turtles are another widely-studied marine animal that is suffering negatively from marine plastic pollution. I remember a few years ago visiting the Mon Repos loggerhead turtle nesting site on Queensland's tropical coast. These highly mobile sea-going turtles often spend months touring the oceans, following the ocean currents and tracking their food. Studies have shown that juvenile loggerhead turtles can spend up to 10 years at sea, swimming an incredible 13,000 kilometres before returning to their hatching place to lay their own eggs and continue the species. Loggerhead turtles have an ability to track themselves back to their birthplace, with seemingly very little effort. These fascinating creatures have long been a mystery to scientists, but it is now believed that they find their way based on geo-magnetic signals, using Earth's magnetic field almost like an in-built global positioning system.

Long before beating plastic pollution became the latest fashion trend, I recall listening to the dedicated rangers and volunteers at Mon Repos explain the two major threats to the survival of the loggerhead turtle population: climate change and marine debris. These remarkable creatures nest slightly above the high-tide line on sandy coastal beaches, like those of the Queensland and Florida (USA) coasts, and their nests must remain dry. Changes to global temperatures and the subsequent rising of sea levels, in conjunction with intensified storm surge events, means that their nests may soon become inundated with sea water, rendering the eggs unproductive. An additional challenge is that the temperature of eggs in the nest determines the gender of the hatchling: too hot and all the hatchlings will be female, too cold and all the hatchlings will be male. So the loggerhead turtles are highly susceptible to the impacts of global climate change.

The other great challenge the loggerhead turtles face is marine debris. Marine debris can take many forms, but most commonly the loggerhead turtle is impacted by fishing netting and floating marine plastics. Due to their size, loggerhead turtles are often accidentally caught as by-catch in commercial fishing nets or they become tangled in ghost fishing lines and netting. Ghost lines and nets are those that have been lost in the ocean by fishing vessels and ultimately float around as marine debris. The plight of sea turtles has been well documented online with various social media videos showing marine conservationists extracting straws and other plastic debris from the nasal cavities of the sea turtles. Unlike seabirds, sea turtles appear to ingest larger fragments of marine plastics. A study of loggerhead turtles in the North Atlantic Ocean showed that of the 24 juveniles examined, 20 had ingested marine debris which was composed entirely of plastic.[10] Of the debris, 25% was large fragment plastics.

Researchers investigating sea turtles in the Pacific also speculated that a range of chemicals may be entering into the turtle's fatty tissue as a result of their ingesting plastic. While the study did not conclusively show a link between plastic ingestion and the accumulation of chemicals in fatty tissue, it did show a possibility that chemicals were being transferred from the plastics into the turtles.[11] The idea of plastic pollution as a vector for exposure to other contaminants will be explored later in this book, but for now, let us just stop and think about the loggerhead turtles: their unique ability to spend years as sea, swimming thousands of kilometres before finally finding their way back to their birthplace; an instinctive repertoire that has occurred for thousands of years. They are highly sophisticated and developed creatures, and yet powerless to withstand the onslaught of humans and their disregard for the natural environment.

Just when we thought we had heard it all and that there could not possibly be another example of the scourge of plastic pollution in the marine environment, work conducted by a leading Australian research team has drawn our attention to a new plastic pollution problem. In 2019, a new study calculated the rate of entrapment of hermit crabs in plastic debris on Henderson Island and Cocos (Keeling) Island.[12] The study showed that, depending on the island and the pollution loading, between 61,000 and 508,000 crabs are estimated to become entrapped in plastic debris and die each year.

You might be thinking 'so what?' when it comes to hermit crabs. Many of us remember having a hermit crab as a pet in our childhood. These 'novice' pets are the perfect intro-duction to 'real' pets, and if the kids can't learn to look after something as simple as a hermit crab, then they definitely are not getting a puppy. Such is the attitude to hermit crabs because of our relatively shallow understanding of their function in the ecosystem. As a starting point, hermit crabs provide an important food source for many marine species including fish and birds. Hermit crabs also perform important ecosystem services in the sense that they act to bioturbate the sand at the beach, in the dune fields and in the intertidal estuary flats. 'Bioturbate' means that they dig through the sand and transport nutrients into the deeper sands. Hermit crabs really are one of those animals that we often look at and dismiss without considering their true importance in the world around us.

Sadly, however, the study alludes to the reality that hermit crabs are threatened by the ever-increasing load of plastic pollution encroaching on their homes. The most notable thing about this study was their finding that the plastic debris threatening these hermit crabs is derived from a very distinct source – plastic bottles. As we will see in the next chapter, plastic drinking bottles are a global menace. They are available in almost every market in the world and have become as globally ubiquitous as coffee and tea. It stands to reason that they are also the most common plastic debris found on global beaches, serving to impact the local marine ecosystems. The source of this plastic pollution, as it was discovered, was primarily ships following major global trade routes in increasing numbers[13] – another example of the compounding impact humans have on the natural environment.

Environmental Longevity of Plastics and Their Intergenerational Impacts

The biggest issue with plastic pollution is its longevity in the environment. We explored earlier the Mariana Trench and the recently documented plastic pollution in the deepest part of our oceans. I started to wonder how much of the plastic that enters the oceans actually settles to those depths. I stumbled across another study that looked at the fate of plastics that had entered the oceans since 1950, and it showed that around 99.8% of plastics that enter the ocean settle to depth.[14] This translates to a massive 8.5 million tonnes of plastic every year. I tried to find a useful reference for this enormous amount of waste to get my head around just how much this was, but the only thing that I could find was an offshore oil platform in Newfoundland that was only about one eighth of that weight.

You may think that it takes many years for the plastic discarded by humans to reach those ocean depths, but you, like me, must be shocked to learn just how quickly humans have managed to pollute the most inaccessible depth of our oceans. Scientists who have examined other deep-sea trenches are beginning to show the impact that the accumulation of plastic is having on the animal popu-lations in these areas. At the Rockall Trench in the North Atlantic, bottom-dwelling (benthic) organisms have been examined for their ingested microplastic composition.[15]

The organisms were all collected from 2,000 metres deep. Incredibly, specimens collected over a 39-year period from 1976 to 2015 were examined. During that period, 45% of the organisms had ingested and retained microplastics, made from eight different types of plastic. The authors of that study, rather interestingly, noted that there was a change in the types of plastic ingested over the study years, with the more recently sampled organisms containing more of the low-density polystyrene type plastics. This type of plastic is typically used as an insulator in building and packaging.

Helping Animals Survive Plastic Pollution

There is no doubt that the natural world is facing a huge challenge fending off the impacts of plastic pollution. The evolutionary sequence is far too slow to allow organisms to comprehend and subsequently adapt to the impacts of plastic pollution. Already we are beginning to see the loss of biodiversity as a result of plastic pollution and this trend is only likely to continue. Alarmingly, the visible impact on animals is simply the tip of the iceberg. Many of the impacts of plastic pollution are occurring at a scale that we cannot see with the unaided eye.

CHAPTER SEVEN

Miniscule Plastic

On a mid-summer day in London, 2017, the outside temperature had reached 30°C. The Tube must have been almost 40°C and commuters were beginning to suffer from the hot conditions. While in many parts of the world this temperature may not seem extreme, in the United Kingdom, outside temperatures almost never exceed 30°C. The last thing that needs to happen on a day like this is commuters collapsing on the platform, or worse, on the Tube itself. On the station concourse there stood an attendant, who was handing something to the bustling commuters to ease their pain: a plastic bottle filled with 350 millilitres of water. This plastic bottle boasted on the label to be the UK's freshest drinking water. The commuters grabbed the bottles with gusto and gulped the water down in seconds. What happened to the bottle then? Was the bottle kept by the consumer until it could be recycled or reused, or was it tossed away – maybe in the bin if one was to be found, maybe to the pavement of the concourse – to become somebody else's problem? Once again, the challenge and question of how we manage plastic as a resource rears its head. Those fleeting 350 millilitres of water have now turned into thousands of years of plastic pollution and waste management challenges. Perhaps a little forethought on behalf of the commuters could have resulted in them carrying a reusable water bottle from home, one they could fill during their day from public refill points, to prevent the production of more plastic pollution.

Aside from the extensive burden that bottled water places on our global environments and the waste stream it generates, bottled water often harbours some secrets. Many product labels will have you believe their bottled water is derived from the most natural of springs hidden high in a rainforest and that the greatness of the natural world has been captured in 350-millilitre increments for your ultimate rehydration. In truth, it is often simply tap water filtered and bottled in industrial estates in the backlots of suburbia. Estimates suggest that consumers are paying upwards of 1,800 times the price for the same water that they could get out of the tap at home, or free at public refill points found in most cities throughout the world.[1]

In many parts of the world, drinking water taps are provided in public spaces – for free. In Rome, for example, the idea of free public drinking water fountains actually dates back to the days of the ancient Romans. Scattered across the city are cast iron standpipes called *nasoni* and wall-mounted outlets called *fontanelle*. These water points provide clean, fresh drinking water, piped to the city from a spring in the mountains high above, to the residents of Rome. There is even a map and smartphone application that allows you to find the nearest nasoni or fontanelle.[2] Simple take your reusable bottle the fountain, fill it up and enjoy!

Many consumers cite the taste or even the health benefits of bottled water as their reason for consuming it.[3] At the end of the day, bottled water is a very expensive way to drink what is a very cheap resource in most parts of the world. But perhaps I'm being too harsh on bottled water. To be fair, many bottled waters are more than just water: they contain a range of trace minerals that may help to balance the body – or so we are told by clever marketing teams.[4]

Recent research from 2018 has also shown that many bottled waters contain plastics. In one study of 259 bottled waters from nine countries, 93% of bottles tested contained plastic.[5] You might be wondering why you have never seen plastic floating around in your bottled water. Well, the plastics that bottled water contain are usually microplastics – micrometre-sized fragments of plastic. The scientists involved with the 2018 study had to use a specialised technique of 'tagging' the micrometre-sized plastic particles with Nile red (NR) dye and visualising the microplastic particles using a Fourier Transform Infrared Spectrometer, a machine that fluoresced (made glow when exposed to light) the NR dye that had bonded to the plastic particles. In this way, the scientists were able to determine an average bottled water microplastic load (of plastic 6.5 to 100 μm in size) of 325 microplastics per litre. To give you an idea of a comparative size scale for these microplastics, a human hair has an average thickness of 70 μm. This gives us a little bit of an understanding of *why* we can't see the microplastics in our bottle of water.

Microplastics

Microplastics have become the latest hot topic for scientific research. A simple Google Scholar search of the term 'microplastic' returns a massive 1.4 million hits in the last five years. Microplastics, it would seem, are all around us. While the exact definition of microplastics is widely debated, they are generally recognised as being plastic particles that are smaller than five millimetres in diameter. Little wonder you can't see them floating around in your bottled water.

Microplastics have been found in beach sand, river sediment and even agricultural soils. Typically, microplastics are the break-down components of larger plastic particles that have degraded in the environment by physical or chemical actions. Microplastics are pervasive and ubiquitous in the global environment. While we do not currently have an estimate for the number of microplastics present on the terrestrial Earth, we do have an approximation of the number of microplastics that are present in the deep ocean.

In October 2020, the Australian CSIRO published work in *Frontiers in Marine Science* that estimated 14 million tonnes of microplastics on the deep ocean floor.[6] A slightly earlier study, published in August 2020 in *Nature Communications* by the National Oceanography Centre in the UK, estimated that in the upper 200 metres of the Atlantic Ocean up to 21.1 million tonnes of microplastics are present.[7] If that number sounds huge, then consider this. The estimate provided is for only a small fraction of the plastic classes, and is constrained to the 32–651 μm size. We could easily expect the estimate to double or even triple in value if *all* microplastics were considered. It is difficult to fathom just how big of a contributor microplastic pollution is to the global plastic pollution budget. It stands to reason that we are therefore discovering microplastics in more diverse and unique locations with each new research study.

Just as you think you are getting a handle on all the possible places in your life that plastics might rear their head, you look at the latest news headlines and realise just how wrong you are. As if it wasn't enough for microplastics to invade your drinking water or the sand on the beach, they are now cropping up in our table condiments. Yes, that's right, research is now showing that microplastics are being detected in table salt. A study published in *Nature* in 2017 showed that in 17 brands of salt, there was between one and ten microplastics per kilogram of salt. That equates to approximately 27 particles ingested per person per year.[8] A more recent study that used microplastics in salt as an indicator for the distribution of microplastics in the world's oceans showed a microplastic content of the salt up to 13,629 microplas-tics per kilogram of salt.[9] Much of this salt is extracted by evaporation of seawater, so when we have estimates of microplastic loads in seawater in the millions of tonnes, it makes sense in hindsight that salt should also contain microplastics. So here we are, applying salt to our evening meal – our evening meal that contains microplastics in the fillet of fish, plastic compounds in the vegetables and microplastics in the salt – all washed down with a nice bottle of microplastic water.

Inside Our Bodies ...

What happens when microplastics make their way into our bodies, and where do they go once they're inside? The truth is we have a very scant understanding of the impacts that the ingestion of microplastics could have on the human body. We simply don't have the analytical tools or techniques that would allow us to confidently detect and then assess the impact of the chemicals and compounds that may be leached into the body from microplastics. What we do know is that microplastics are not fully retained in the body. Recently, a small pilot study was presented at the United European Gastroenterology conference that studied the fate of ingested plastics in the body.[10] The study found that microplastics are passed through the human digestive tract and excreted in faeces.

Although only eight people took part in the pilot study, the researchers found microplastics in all of the samples. Unfortunately, there was no quantification of mass balance between microplastics going in versus those going out, so it is difficult to speculate about the retention capacity of the body in this instance. The passage of microplastics through the body makes sense though, when you stop and think about it. I remember as a child being reassured by my parents when I accidentally swallowed a cherry seed that it would end up coming out the other end the next day and to not worry about it. That is the human body doing exactly what it should do: discarding the waste that it can do nothing with.

Humans are highly mobile animals. Like birds and sea turtles, humans get around. Humans are at home in the morning, then commute to work, then go to the gym, then to the shops, then back to work, then back home again. Somewhere in that daily schedule, nature calls. So that bottle of microplastic water you drank on the train station platform on the way to work, combined with the microplastic salt that you added to your grandmother's bland boiled vegetables the night before, and the microplastic chicken pie that sat alongside those bland vegetables, all have to go somewhere. Somewhere in your day, you liberate yourself of the microplastic excrement. Just like nature intended, you rid yourself of the unwanted material in your body. Meanwhile, nature's great mechanism for ensuring species diversity, abundance, and dispersal means that through the act of going about your daily life and using the bathroom, you have transported those microplastics from their origin in the form of food and beverages, into your body as you consume them, then out into the world somewhere else. Just like the great African elephants on the savannahs of Africa disperse seeds, the human body is passively acting as a dispersal mechanism for plastic pollution throughout the global environment.

Now we could stop talking about plastic-laden excrement, but where would be the fun in that? While reading the online news an article caught my eye. It was one of those 'click bait' headlines, and true to the name, I was baited and clicked on the headline. The headline read something like 'plastic fibres found for the first time in wild animals' stool'. Now, doesn't that sound like a terrific article to dig into deeper? Well, I wasn't disappointed; the article detailed how a team of marine scientists had identified plastic fibres in the stool of fur seals, and they were using the stools as a means for tracing plastic pollution through the oceans.

Scatology

Scatology is the study of excrement, otherwise known as stool or scat, in order to understand something about the organism that produced it. It normally focuses on the dietary habits and health of the organism. It makes sense, then, that we should consider the output of animals in order to understand how plastic pollution is distributed around the world, but also, and possibly more importantly, how plastic pollution finds its way through various organisms.

The study of fur seals, published in *Marine Pollution Bulletin,* showed that 67% of the 51 South American fur seals studied contained microplastic fibres in their scats.[11] The study was conducted in the northern Chilean Patagonia. This was really a game-changing finding. No longer are we left hypothesising that animals may act as vectors for plastic pollution and redistribute the plastic pollution throughout the world, we now have the evidence to show that is actually happening.

As is usually the case in science, there is often a rapid onset of studies that show very similar findings once one study has 'broken' the story. As the media was writing about the South American fur seals, another study was also being published in the *Marine Pollution Bulletin* looking at the northern fur seal.[12] Although they didn't look for the same size fraction in the scats as the South American fur seal study, this second study found microplastic particles in 55% of the 44 fur seal scats examined. Between the two studies, we are really starting to get a startling picture of the reality of plastic pollution and its pervasive behaviour in the global ecosystem. These two studies show us that more than half of the indicator species examined, the fur seals, contain plastic in their scats.

Let's pause for a moment and think about how the plastic actually got there. The fur seals didn't just suddenly accumulate plastic in their bodies. It means that plastic pollution had to be in either their living environment or their food. It means that the seals then had to ingest and digest those plastic particles, which then were either broken down and absorbed into the seal, or excreted. The microplastics were an integral part the ecosystem and living environment of those seals. The fur seals, in this example acting as the apex predators, are ringing the alarm bells that microplastics are abundant in this environment and in this food chain. What further evidence do we need that shows how deeply intrusive humans have been into the natural processes of the natural world?

Mosquitos

When we thought we knew a lot about microplastics, it turns out we really knew very little. How we thought organisms interact with microplastics in the environment, in a way that could be described as childishly simplistic, has recently been turned on its head. Scientists have been studying the way microplastics are internalised into organisms and have made some startling discoveries.

We can all relate to the annoyance of mosquitos – well known for their ability to transfer tropical diseases through an environment with ease. In many parts of the world, mosquitoes and the diseases they act as a vector for have been responsible for mass declines in human populations. Evidence now also points to mosquitoes as vectors of transfer for microplastic pollution. Mosquitoes seem to be able to transfer tiny fragments of plastic between animals as they move from individual to individual.[13] While the link between mosquitoes and the transfer of plastics between humans hasn't yet been made, this study points to an alarming possibility. But there is also an unanswered question that the discovery poses – how *else* do microplastic enter the body of an organism?

Up until a study was released in the December 2020 edition of *Science Advances*, scientists were really unsure of the answer to this question. Researchers had some ideas, but the supporting evidence wasn't yet strong enough to draw a conclusion. The study, undertaken by a team at Bayreuth University, showed that microplastics are drawn from the gastrointestinal tract of aquatic animals, aided by specialist bacteria and microorganisms, into tissues by a process known a cellular internalisation.[14] The microorganism-coated plastic particles become preferentially absorbed into body tissues as they are recognised as a source of sustenance. Essentially, the microplastic particles are being masked and are fooling the aquatic animal into thinking that it is a source of food. Once internalised, the microplastics can then make their way around the body of the animal with ease. The implications of this are of course enormous, but there is still no strong evidence to indicate that humans internalise plastic pollution in the same way.

In 2022 the concern of many researchers was finally realised: plastic pollution has the ability to enter the *human* body. Not just in the intestinal tract, but into the blood. In a particularly significant discovery, scientists examined the blood of 22 study participants, of which 17 had detectable concentrations of plastic in their blood.[15] The question of *how* the plastic entered the blood was the focus of some debate until a little bit later in 2022, when another group of scientists discovered plastic in human lung tissue, seemingly providing a plausible answer. In this study, researchers examined live lung tissue removed from patients during surgery. Of those patients, 11 out of 13 had microplastics present in their lung tissue.[16] The researchers commented that a big concern is exactly where in the lungs the plastic particles were found – deep in the lungs where those types of particles should not be able to reach. With so little information available to medical professionals about the impact of plastics on lung tissue health, the researchers raised concerns that the sharp rigid particles of plastic may cause the same damage as asbestos fibres, suggesting that plastic pollution may be the source of the next lung disease epidemic.

No longer is the conversation around the impact of plastic pollution on humans one of hypotheticals and speculation. The jury is in and the evidence extremely compelling. We are, without a doubt, poisoning ourselves with plastic pollution.

CHAPTER EIGHT

Plastic Pollution at Extremes

Plastic in the Sky

We've all been there, sitting at 35,000 feet watching our in-flight movie, loathing how the ever-encroaching passenger sitting on the next seat has managed to envelope the arm rest and most of your seat while snoring at a volume well above that of the roaring jet engines. Don't look at your watch again, those 12 hours aren't going any faster. To add to the excitement, the meal cart has just appeared. The smell of the freshly heated food wafts through the cabin. The chicken or the fish? A bountiful dose of microplastics, or just a little one? Most people probably don't think about this as the cart trundles along the exhausted rows of passengers, but the more eco-minded know exactly what is coming down the aisle on that trolley. Aside from the horrendous stomach-aches and an emergency dash to the lavatory, that cart also means a litany of plastic packaging, wrapping, cups, cutlery – a plastiphobe's nightmare. Each item on the food tray meticulously wrapped, each bread roll, each slice of apple, all presented in a 'convenient' and 'hygienic' single-use package.

Stop there! Don't even bother to offer up your reusable water bottle to the flight attendant. They will just stare, bemused, back at you until you retract the offending bottle that is now receiving glares from fellow passengers for holding up the meal delivery service. You have no choice, it's a plastic cup or dehydration. Well, at least you can use the plastic cup next time, right?

Flying is one of modern human's greatest accomplishments. With the aid of jet engines (fuelled using fossil fuels) we are capable of travelling huge distances in relatively short times. The altitude at which we fly and the delicate conditions that we must mimic while flying all make air travel an extreme environment in which we are finding sources and sinks of plastic pollution.

While many airlines are now looking to reduce their reliance on single-use plastics, many are not. Often, claims are made regarding the efficacy of keeping food stored and hygienic without the aid of single-use plastic packaging. Plastic does have its place in the world, and like so many great inventions, was once used with the greatest of intentions, but how can we now justify the wrapping of every item on a plane meal tray in single-use plastics when we know the damage it is causing to our global environment when not managed correctly? In 2018, it was estimated that approximately 4.3 billion passengers would have flown on a commercial airline in just that year.[1] Admittedly, not all of those passengers would have been on flights where a meal was served; and on flights where a meal was served, not all meal items would necessarily have been individually wrapped in single-use plastic. But even if 50% of the global airline passengers in 2018 consumed just one plane meal in their journey, with each meal containing two items of single-use plastic packaging, that means that 8.6 billion pieces of plastic packing have been used and discarded, 35,000 feet above our heads. Add to that water bottles, snacks, plastic packaging around blankets and headphones, and we have a massive amount of plastic waste. Air travel is possibly one of the biggest generators of single-use plastic waste in the modern world.

Some airlines combat the waste problem by incinerating their single-use plastics, while others send it away to landfill. Regardless of their method of disposal, there is still an environmental burden generated by these single-use plastics by the gases produced during incineration or the persistence of the pollution by way of debris or leachate when sent to landfill. Really, the only solution to disposing of the plastic waste is to not generate it at all.

This was exactly what Qantas, Australia's national airline, did in May 2019.[2] In a world first, Qantas operated a zero-waste flight from Sydney to Adelaide. On this flight everything was either recycled, composted or reused. Instead of the usual 34 kilograms of waste directed to landfill, Qantas managed to divert all waste and create a zero-waste footprint for that flight.

Excitingly, zero-waste in aviation is not just limited to the skies. San Francisco International Airport (SFO) is one of many around the world that are taking substantial measures to reduce the burden created by the waste generated through travel.[3] By 2021, SFO was expecting to become a zero-waste airport. This means that at least 90% of the waste created at the airport would be diverted from landfill into other streams like recycling and composting. Their plan also included a food recovery program that would divert surplus food from landfill into a meals program, providing around 13,000 meals per year for local community groups and charities.

While travelling, you can have an immense impact on your plastic waste burden just by being mindful of a few simple things: use a reusable bottle where you can, use a reusable coffee cup where you can, avoid taking food you don't need, and finally, do some research about the airlines you use and the airports you visit. It is remarkable just how many international airline hubs now cater for zero-waste and reduced-plastic travellers – you just have to look!

Everest

Descending a tiny bit lower than the cruising altitude of our plastic-filled plane, we begin to approach the top of the highest peaks on Earth. Mount Everest, in the Himalayas, is the highest. At just under 9,000 metres, Mount Everest is only slightly shorter than the 10,000-metre cruising altitude of a long-haul plane. For thousands of years, humans and animals have gazed at the peaks of the Himalayas wondering what secrets the snow-capped summit holds. The mountains have been central to many religious beliefs of the region, with the local indigenous people, the Sherpa, considering the mountains to be sacred. The unique form of Buddhism practiced by the Sherpa people means that they have the greatest respect for the mountain that they affectionately refer to as 'Mother of the World'.

Despite having inhabited the region for generations, the Sherpa people did not see a need to summit the great peak of Mount Everest. This burning desire was realised by the explorers and adventurers Sir Edmund Hillary and Tenzing Norgay in 1953. Since that day, adventurers and mountain climbers have all dreamt about climbing and conquering the great peaks of Mount Everest. Every year, hundreds of tourists set out to climb this iconic and sacred mountain. While some reach the summit, many discover that they aren't actually as fit as they thought they were and soon settle for a slightly less arduous journey to one of the many check-points along the climb. At these check-points, some walkers decide that they've been carrying too much food and excess weight and so decide to discard it. Some climbers discover that they should have paid that little bit of extra cash to get better-quality walking poles as theirs have now bent in half, so they leave them on the side of the track. All manner of waste is simply discarded on and off to the side of the track. Mount Everest has now become a waste dump. While efforts are made at the close of the climbing season to clean up some of the waste, it is too big of a challenge to transport the thousands of kilograms of waste back down the mountain to a place where it can be handled appropriately. The upper peak of the mountain, the 'dead zone', is one of the most polluted parts of the mountain. Recently, eight tonnes of rubbish was removed from the dead zone, but it is believed that almost 50 tonnes still remain there.[4]

A study was conducted in 2019–2020 that examined plastic pollution in both snow and stream waters on the mountain.[5] The study found that in the snow there was approximately 30 microplastic particles per litre of snow melt water. By comparison, the stream waters contained a much lower concentration of microplastic particles: one microplastic particle per litre. These microplastic particles were comprised almost entirely of polyester fibres that the authors of the study attributed to the polyester (i.e., plastic) based clothing used by tourists on the mountain. Incredibly, microplastic fibres were detected as high above sea level as 8,440 metres, on the Balcony – one of the highest 'easily' accessible places for the general tourist to reach on Mount Everest.

Various strategies to combat the pollution problem have been introduced, including a weigh-in/weigh-out procedure for climbers and a waste levy, however many climbers don't seem to understand the issue and continue to discard their waste into this once-pristine extreme environment.

Fuji

During a work visit to Japan a few years ago, I made a trip to another of the great mountains – Mount Fuji. The mountain itself is actually a volcanic cone, a geologically active volcano located not too far from the biggest city of Japan, Tokyo. In keeping with the Japanese tradition of respecting the environment, Mount Fuji and the surrounding region was spotlessly clean. This is despite the high tourist traffic that visits this iconic volcanic feature. It is a place of immense beauty and the power of the landscape can be sensed with every breath of the wind. There is little wonder that this mountain, like Mount Everest, has been a cultural and religious feature for many thousands of years. Although there was no evidence of plastic pollution around the main walking trails, there were a number of cleaners tasked with monitoring the trails to ensure that they remained void of the remnants of human visitation.

In what struck me as a strange juxtaposition with the pristine and spotless walking tracks, a gift shop was placed strategically along the walking path to one of the lower check-points. The gift shop had the regular curios including stuffed toys and fridge magnets, all emblazoned with the words Mount Fuji and a picture of the distinct cone shape of the volcano. What stood out to me as both remarkable and troubling was that every single item in that gift shop was wrapped in or contained at least one piece of single-use plastic packaging. Why, in this otherwise pristine environment, tourists are encouraged to purchase products that are packaged with single-use plastics I'll never fully understand, but I was glad to see that the plastic wasn't being discarded haphazardly around in the environment.

Plastic-Conscious Travel

We are really just starting to understand the full extent of plastic pollution on Earth, but the evidence that shows us plastic pollution extends beyond the common domains of rivers, soils and oceans is profound. Fortunately, we can do two simple things to help reduce the plastic pollution burden that we have on the planet when travelling to remote and extreme environments. The first is taking out what we take in – the environment is not a dumping ground and we should all respect wilderness areas and places of pristine natural beauty. If we can carry the packaging in with us, then we can carry it back out again. The second is being mindful of the resources we use when in wilderness areas. Instead of polyester clothing, consider natural fibres like wool, hemp or bamboo, all of which have properties that often exceed those of plastic-derived textiles.

It is also important to select quality. Rather than purchasing resources that are cheap and will likely break, spend a bit more money and get sturdy, long-lasting clothing and equipment. Purchasing gear that will last longer and not fall apart will reduce the number of particles and debris that is liberated into the environment as it degrades. We really don't have to change our entire lives to continue to enjoy the Earth in a reduced-plastic, environmentally-friendly way; we just have to stop, think and plan our journeys so that our only remaining trace is a footstep.

CHAPTER NINE

Space

When we talk about global plastic pollution, we generally think, as we have done until now in this book, about the oceans, animals, and to a lesser extent the land surfaces. If the depths of the oceans are one of human's last frontiers of endeavour, then space is the other. Humans have a desire to explore the universe around us, to understand the unknown and to explore far-off galaxies in hope of finding something special – perhaps life on another planet.

We are all familiar with the footage of spacecraft departing their launching stations around the world as they project themselves through the atmosphere and into the great vastness beyond. We know that spacecraft are a source of fossil fuel pollution as we see the smoke released from the huge engine cones on launch. We are also familiar with the idea of spacecraft creating substantial debris, because sometimes things go wrong and we see on the evening news footage of investigators collecting pieces of the spacecraft in an attempt to figure out what exactly happened. Generally, we understand fairly well how spacecraft can act as a source of pollution in our environment. But how many people stop and consider the pollution generated by spacecraft when they are beyond Earth's atmosphere? How many people realise that the human endeavour to explore space is also turning near-Earth space into a dumping ground?

In 2021, one news outlet reported that more than 27,000 objects larger than a cricket ball were travelling around in Earth's orbit, as what we colloquially call space junk.[1] This space junk is a host of components from all manner of space-bound devices including space shuttles and satellites. What many people are unaware of is that these devices are furnished with plastic-based materials and high-performance polymers specifically designed for space-craft and use in extreme environments. DuPont's Vespel polyimide products, for example, have been specifically designed to be used in extremely low-temperature environ-ments and where a vacuum (such as in space) may result in regular plastics producing too much friction and wear. There is a range of legitimate and important applications for plastics. Thermotolerance for applications in space travel is, without a suitable alternative, one of those instances when it makes sense to use plastics. It is not the plastic itself that presents the problem, but rather the way in which these plastics are managed when their utility has expired.

It really does make startling sense: humans can't control their consumption and pollution on Earth, so why should it be any different in space? Thanks to space exploration, satellites, telescopes and all manner of other things launched through the atmosphere, Earth now has a nice orbiting mass of pollution. If there is any life in the galaxies beyond our own, then this great universal garbage patch should act a warning sign to visiting life forms of what is to come should they penetrate Earth's atmosphere. While the pollution orbiting Earth doesn't necessarily resemble the common plastic bottles and straws that litter the global beaches, choke the global rivers and are found at the deepest depths of the oceans, the orbiting pollution still contains many of the key components that make plastic pollution on Earth so problematic. Astoundingly, there is a body of thought that space may be a suitable dumping ground for even more waste.

In 2016, BBC Radio 4 broadcast an episode of *The Curious Cases of Rutherford and Fry* titled 'The Stellar Dustbin'. The discussion was focussed around the possibility of discarding rubbish by launching it at the sun. Great idea, right? Well, as it turns out somebody has already thought about this and decided that at a cost of £29,000 per kilogram, it was not a financially sensible undertaking. It is really interesting that the narrative around the topic of using space as a rubbish dump is much less of 'should we and could we', and more about whether it would be a financially viable exercise. In years to come when space travel inevitably becomes cheaper, will it also become cheaper to launch rubbish into space? For the sake of the universe, we can only hope that it does not.

Space junk is actually a very complex problem and is not a new conundrum. It has been a fundamental problem associated with the human endeavour of space travel since the very first spacecraft left Earth. For this reason, space junk has featured as a key component of the *United Nations Treaties and Principles on Outer Space*, with Article IX declaring:

> 'States Parties to the Treaty shall pursue studies of outer space, including the Moon and other celestial bodies, and conduct exploration of them so as to **avoid their harmful contamination** and also adverse changes in the environment of the Earth resulting from the introduction of extraterrestrial matter and, where necessary, shall adopt appropriate measures for this purpose.'[2] [emphasis added]

Despite the best efforts of the United Nations and their treaties, space junk has become a serious problem that will require critical thought over the next decades. The threat was seen in 2015, when space junk caused crew on the International Space Station to seek emergency shelter after a piece of debris from a former weather satellite threated the safety of the crew. Again in 2020 (Expedition 63), NASA was required to take drastic actions to avoid a collision with space junk floating within dangerous distances of the International Space Station.[3] As we become more dependent on satellites and remote technologies, more devices will have to be launched into space. More and more obsolete components will be discarded at their end of life as 'junk'. The problem of space junk has become so big that there has been commercial interest in cleaning it up, with the value of the pollution components estimated to be around US$300 billion.[4] There may also be interest from traditional fields of academia that are searching for new non-traditional research avenues. In a paper published in the journal *Antiquity* in 2021, archaeology researchers proposed a method for contemporary archaeological work associated with space exploration and launched the International Space Station Archaeological Project.[5] It is not a stretch of the imagination to think about researchers recovering space junk and unravelling space exploration stories from times gone by.

The pollution problem in space is so significant that there is a global team of scientists who monitor, using lasers, the trajectory of the pollution in orbit to ensure that it doesn't bump into active satellites. Could you image the global chaos that would unfold if we lost some of our satellite television capabilities for a day? Or if our in-car GPS receivers couldn't tell us to 'continue straight for 500 metres'? Or if you were trying to navigate a simple journey without mobile phone-based GPS navigation? Our reliance on technology has meant that the ability to read a simple map is a skill lost to many.

The 'Invisibility' of Plastic in Space

So much of the attention given to plastic pollution is focussed on the plastics that we can see. Those that we can't see are often overshadowed. Plastic pollution in space is currently overlooked by the broader community because it is, to those observers, invisible. We have become well accustomed to spotting those plastics that stand out to us as we walk along a footpath, a river, or the beach, and are distributed as regularly as leaves and flowers in shrubs or as regularly as seashells on the beach. These plastics are often bottles, bags or packaging. These plastics are by most accounts the easiest to clean up, and most importantly for many, the most social-media friendly. They are large and when removed, the resulting visual is impactful.

The challenge that space junk poses is that, for most people, it is not something that is ever seen. It is not something that sits in our line of vision when we go to the beach, river or park. For most people, reading this book is probably the first time that space junk has ever registered as a problem. The unseen is very often the unchanged. Space junk is not glamorous, most people will never have the opportunity to touch it, and it certainly doesn't create social-media-worthy content. However, it is something that all countries need to manage and the sooner strategies are put in place to manage end-of-life plastic resources in space, the less of an impact humans will have on environments beyond our own planet. This can only happen if committed individuals start the conversation and seek change. Hopefully, this section of the book inspires *you* to take this on as your contribution to the Plasticology Project.

CHAPTER TEN

Plastics in Soil

We have become accustomed to the social media videos and evening news grabs of plastic in our oceans. We know that the global oceans and rivers are in dire need of a plastic pollution overhaul. But what about the plastic pollution in global soil?

Soil plastic pollution is often overlooked by plastic pollution campaigns for a number of reasons; however, the primary reason is the relative lack of understanding about the global soil environment. Despite being the place where we build our homes and plant our crops, there is an overall general disconnect between many of us and the soil beneath our feet. Why is this the case, though?

I remember studying soil science as an undergraduate student, learning about the soil horizons (layers of the soil) and what they meant in terms of the organic matter of the soil, the age of the soil and formation processes responsible for creating that soil landscape. I remember being shown the different colours of soils, using a colour chart to interpret various parameters about soil origin, and using sieves to separate particles of different grain sizes in order to examine them under a microscope. This was all to understand how soils moved and developed in the landscape. I also remember very clearly one of my lecturers telling me that 'sometimes the only way to know what a soil is made from is to taste it'. Yes, that's right, I sampled the culinary delicacies of the Australian soil landscape. But while all of this is useful information from the perspective of a soil scientist, it isn't intrinsically interesting to the general population. People generally don't find the idea of digging a hole just to look at the changing colours and then snack on a clod of the orange layer particularly appealing.

Bioturbation

So let us actually explore the soil environment. Recently I had the fortune of travelling for work to Namibia, a country in southwest Africa. It is a beautiful country, with a very small population compared to its landmass, and little development outside of the main urban centres. Conservation seemed to be at the heart of the community everywhere I travelled.

Although fragmented into large game reserves, the wilderness areas are a safe haven for iconic African animals including African elephants, giraffe, antelope, and warthog. The African soil plays a vital role in the functioning of the entire ecosystem. Below my feet I would see the constant movement of highly industrious ant colonies, turning over the soil – bioturbating – as they build a colossal network of tunnels below the soil surface. Squirrels burrow their homes in the surface soil and then scramble about playing peak-a-boo through their labyrinth of holes. Warthogs forage with their snouts and tusks in the grasses and surface soil, searching for that most tasty of food morsels. The great African elephants scoop huge trunkfuls of soil and toss it over themselves after they've spent a good long while washing in the waterhole. The soil is said to soothe the skin and relieve irritation from the sun.

While all of this activity of turning over the soil is occurring, there is an unfortunate and unplanned outcome. Anything that sits on the surface of the soil is worked down into the soil profile. Plastic on the surface, for example, then becomes worked down through the profile and may end up deep within the soil matrix. The plastic pollution therefore no longer only creates a problem when it resides at the surface, but also when it is worked into the soil by the scratching and digging of the bioturbating animals.

The local communities of Namibia also have a strong connection to the soil environment, with their homes often made using grasses and mud-brick, and their food grown in cultivated fields. There is clearly a deep connection to the African soils throughout the country.

One day when visiting a large urban community in the north of the country, I was struck by the amount of plastic pollution that seemed to be in the area. Throughout the town, plastic bottles, bags and packaging were scattered as though they had just been tossed aside by the user and forgotten about. But given the significant connection of the local people to the land, this didn't seem to fit. During my visit to this town, I saw all the structures of the society, from the gated homes with razor wire fences and a guard's watch house, to the unofficial non-permanent housing structures built of tin sheeting and whatever materials were lying around. But in none of these settings were people absently discarding their waste into the environment. If they weren't placing their waste in the municipal collection bins, they were fashioning it into something of utility. The mentality of being wasteful didn't appear to exist. So where was all this plastic pollution coming from?

It wasn't until my final day in this town that I witnessed the source of the pollution. On the edge of the town was a large municipal waste dump, and on each of the corners of the town were smaller stockpiles of waste. Evidently, the municipal waste collection service was dumping the collected waste at these stockpiles on the edges of the town, only to be blown across the town by the strong desert winds. The stockpiles had clearly been used for many years as there was a hillslope formed from compressed layers of soil and plastic debris. On the surface of the hillslope, established trees were growing with roots penetrating down into the plastic layers below. These stockpiles were beginning to act as a natural soil profile, with various horizons forming. Thinking back to those long hours examining soil horizons as a student, I wondered if in future years soil scientists will introduce a new classification to the soil horizon scheme – the plastic horizon.

Waste stockpiles, Namibia.

The Plastic Horizon

The plastic soil horizon has become a reality in many parts of the world. If you have ever travelled in China and gone out of the large cities, you will see that a great majority of the country is composed of small-scale agriculture. This is particularly true of the central and northern regions where the climate and soil conditions are far more favourable for agricultural activities. As you traverse the remarkable Chinese landscape you will see pockets of land being worked by farmers and their beasts of burden. It really is marvellous to see the productivity of these farms and experience the cultural connection of rural China to the soil and landscape. Unfortunately, though, the appreciation of Chinese agriculture is somewhat tainted when you begin to take a closer look at the soils in which the vegetables are growing. It turns out that plastic has been used by farmers for many years as a kind of mulch. Plastic is believed to perform well in maintaining soil moisture, supressing weeds and preventing certain pests.[1] While this use of plastic may have a practical ecosystem function, the tendency for plastics to break down (in both the mechanical and biological form) will ultimately mean that the soil environment is degraded and contaminated.

Critics of the practice have long called for it to be halted, citing the global plastic pollution crisis.[2] The climate of plastic pollution has now changed and as current predictions suggest that around 2 million tonnes of plastic mulch are used every year in China, it has reached a point where authorities also no longer believe this to be the best practice. China is now looking to introduce tighter regulations that will mean less plastic will be permitted to be used in agricultural soils. Multiple English-speaking media outlets in China suggest that the new regulations will require farmers to limit their use of plastic and restrict the type of plastic used to a select few.[3] The idea behind these new, more stringent regulations is to propagate a 'green agriculture' framework in China.

Legislated Plastic Pollution

Demonstrated by the shift in regulatory thinking in China, emerging research is beginning to show the significance of the plastic pollution problem in the soil environment. In Australia, there is a standard that prescribes various conditions for compost, soil conditioners and mulches that are sold for use in backyard gardens and on farms. The standard permits up to 0.5% of rigid plastics in dry matter soils, and 0.05% soft plastics (flexible and film) in dry matter soils.[4] That might not sound like a lot of plastic, but if you do some very simple extrapolatory calculations, you soon realise that for a volume of compost of mulch for a typical suburban backyard, it is possible to have a very large amount of plastic debris and material in that soil. Fortunately for us, scientists have done those calculations and have shown that under the current permissible conditions of plastic debris in soils, when adding 1,000 metric tonnes per hectare of soil, compost, conditioner and mulch over a 10-year period, a one hectare plot of soil could contain up to 5 metric tonnes of rigid plastic debris and material.[5]

Are you having trouble imaging what 5 metric tonnes looks like? Well, think of it this way: a Toyota LandCruiser – one of the largest four-wheel drive vehicles available – is about 3.3 metric tonnes. So in Australia, it is permitted to add the equivalent of 1.5 Toyota LandCruiser vehicles of rigid plastic material in soil, compost, conditioner and mulch to that one hectare area over a 10-year period, in addition to the allocation of soft plastic debris also permitted in the soils, compost, conditioner and mulch for that area.

This permissible soil plastic loading seems inconceivable in the current global climate and understanding of plastic pollution. In Australia, there is a standard operating procedure that permits the inclusion of these pollutants in the soils. We have seen with plastic pollution in other environmental spheres that plastic, in particular plastic that is breaking down, can have huge impacts on ecosystem functioning and health – so why would we permit soils for use in agriculture and home gardening to contain these levels of plastic debris?

The answer is remarkably simple. Globally, we need to recycle our resources in order to combat the problem of too much waste going to landfill, and an eventual shortfall of available resources globally. Building rubble, topsoil, manufacturing debris and other industry waste are commonly 'recycled' into soil for fill, agriculture and home gardens. In addition to these types of waste, we often see the addition of wastewater and sludge to soil conditioners, mulch and fill materials. Much of this sludge is generated as a by-product of the municipal sewerage water recycling process and often contains various particulates, including plastics. The challenge of removing all the plastic debris from that waste material means that it ultimately ends up staying in the end product soils. As a society that depends on intensive commercial agriculture and insists on controlling the environment around us, we have become accustomed to the addition of various conditioners and products to our soils. Adding plastic to soil materials has successfully created another vector for the movement of plastic pollution around our plastic planet – through agriculture and home gardening.

CHAPTER ELEVEN

Plastic Vectors

Have you ever stayed in one of those low-budget backpacker hostels and seen a long line of plastic food preparation boards lining the kitchen benchtop? They're usually presented in a wide range of colours with an accompanying sign explaining what each of the coloured boards should be used for. White is for bread, green is for vegetables, red is for meat and yellow is for chicken – simple. Just don't make the mistake of using the wrong board as you are bound to receive a stern lecture from a fellow traveller who appeared in the kitchen without you noticing and has overseen your mistake. Okay, so we like to chromatically separate our food boards into different food groups, but what is the point?

Whether we realise it or not, plastic is really good at having stuff stick to it. No doubt you noticed that when you used the white bread board to cut a chilli how colour from the chilli seemed to permeate the knife cuts in the plastic, leaving it stained red. In an effort to save yourself from an attack by sleep-deprived fellow travellers, you spend the next 20 minutes hunched over the kitchen sink scrubbing the board clean to erase any evidence of your preparation board faux pas.

In that moment you have learned something about plastic and its unfortunate design limitations. You have learned that smooth, unblemished plastic surfaces are reasonably okay at repelling the assault of stains. Sure, the plastic might retain remnants of food and there may be some surface staining but it is nothing compared to when the board has been well-used and has lacerations and cuts that seem to be an invitation for your incorrect food choice to make a lasting point about your wrong-doing.

This is the very good reason why we separate our preparation boards into food-specific colour groups. We don't want the plastic and the food remnants to be sticking together, and we don't want the remnants of the food's colour pigments, smells, bacteria, or pathogens to remain on the surface and carry over to the next food item prepared on the board. We particularly don't want things like the bacteria from raw chicken to be interacting with fresh bread as this is a direct avenue of pathogen transfer between a high disease-risk food group and a relatively inert food group. We are trying to prevent the plastic preparation boards from acting as vectors for bacteria, pathogens and viruses and essentially rafting these between preparation board users. So rather than just being a matter of décor, the rainbow of plastic preparation boards actually serves a meaningful purpose that is often not given much thought by kitchen users.

Why start this chapter talking about plastic food preparation boards in a hostel? Let's face it, during our younger years, many of us have stayed in some not-so-clean establishments and it wouldn't really come as a surprise to us to find out that other travellers have been using the white board for their red board needs. There is a good chance that even though the bacteria from the meat had mingled with your fresh salad greens, you didn't develop an illness that left you searching the depths of your bags for the anti-diarrhoea tablets. You probably avoided any adverse outcome from the encounter and were none the wiser. But that could be just luck.

Plastic as Rafts

I started this chapter with that anecdote because it shows us that we're already familiar with the idea of plastics acting as vectors – or rafts – of bacteria, pathogens, and other nasty stuff around the environment. Through a simple and commonplace example, we can see that plastics have some role in keeping us healthy or, conversely, making us sick, whether we know it or not. We understand that plastics can be responsible for transporting bacteria from one food item to another. We understand that the implications of poorly managed food preparation surfaces could end in our becoming violently ill in the middle of the night. The point is, we already know that plastic can attract and transport many nasty things. We know that we must exercise caution and adhere to stringent food hygiene practices to avoid spreading bacteria, pathogens and disease around the kitchen – it is well understood and we work with that knowledge accordingly. So why then is it so hard to fathom that plastics in the environment are acting as rafts of bacteria and pathogens around the world? As a global community battling to resolve the issues of plastic pollution, we often overlook the role that plastic plays in the global environment when it comes to spreading bacteria, pathogens and disease. This oversight, as we will see in this chapter, has been in some way responsible for shaping the beginnings of the decade of the 2020s.

So how and why does plastic act as a raft for bacteria, pathogens and other nasty things? We simply do not know enough. This was the conclusion reached by a World Health Organization review into the health impacts of microplastics in drinking water published in August 2019.[1] We simply do not understand well enough how microplastics act as rafts or vectors for metals, organic compounds or even pathogens through the global environment. Although focussed on microplastics in drinking water, the report poignantly pointed out the problem that we have for plastic overall – not enough research into their environmental fate and impacts.

This conclusion seems somewhat surprising on account of the reasonably strong body of evidence that already exists in the published literature regarding the way certain bacteria preferentially bind to certain types of plastics. One such example is of how coral reefs around the world are becoming engulfed by white syndromes as a result of bacteria rafted by plastic pollution. In a study examining 159 coral reefs in Indonesia, Australia, Myanmar and Thailand, it was found that reefs that contained some form of plastic pollution were between 4% and 89% more likely to have disease than reefs without plastic pollution.[2] When the coral was in direct contact with plastic – for example, when a plastic package was draped around the coral – there was an increase of almost 20 times greater incidence of disease in those corals. While the exact reason why this occurs is still not entirely clear, the study found that the plastic wrapped around the coral can cause the coral to stress, deprive it of light and oxygen, and, perhaps critically, the plastic releases toxins into the immediate environment of those corals. What we see is a micro-environment where the healthy corals become unhealthy, pathogens are rafted in and the invading pathogens take hold.

One of the study's authors, Dr Joleah Lamb, noted that 'plastic items such as those commonly made of polypropylene, like bottle caps and toothbrushes, have been shown to become heavily inhabited by bacteria that are associated with a globally devastating group of coral diseases known as white syndromes.'[3] In the study, at the time of publishing, it was predicted that 11.1 billion plastic items were entangled on coral reefs in the Asia-Pacific region alone. That staggering figure is expected to rise by 40% by 2025.[4] What does 11.1 billion actually look like? Well, think of it like this: the Colosseum in Rome could hold somewhere around 50,000 spectators. Assuming that each piece of plastic was one spectator, we would need 220,000 Colosseums to accommodate all of the plastic items estimated to be entangled on coral reefs in the Asia-Pacific. By 2025, we would need 308,000 Colosseums to accommodate the plastic load. Each of these Colosseum spectators is carrying bacteria that can have a deadly impact on the health of the reef.

In another study from 2013 it was discovered that the hydrophobic (water repellent) surfaces of most plastics create ideal colonisation conditions for various bacteria and for the establishment of biofilms – we will consider biofilms shortly.[5] These surfaces can create an abundant ecosystem of hundreds of microbial assemblages that are then transported far and wide throughout the oceans. On that journey around the oceans, the ecosystems may become more diverse, they might see genetic evolution or substantial alteration to their physio-chemical environment that could result in chemical and morphological changes in the plastic particles themselves.[6]

Despite these studies, the World Health Organization's report is very clear in stating that we do not currently have sufficient information available from standardised and controlled investigations to determine what role, if any, microplastics may play in the distribution of plastic bound pollutants across the globe. What an incredibly scary thought that is – around us we have plastic pollutants everywhere, and yet we do not understand what function they are playing in the environment and on global health. This is possibly one of the largest oversights of the plastic era and is akin to other pervasive and catastrophic environmental pollutants such as lead and asbestos. We, as humans, have consistently released pollutants en masse without first understanding their fate in the environment at the sub-visual scale. Only long after the pollutants have left a wave of destruction do we consider their impacts in this way. We have (or so we thought) learned from the ubiquitous use of herbicides in agriculture and their impact on DNA methylation; we have (or so we thought) learned from the prolific release of lead from fuel into the environment and the impact that has on brain functioning; and we have (or so we thought) even learned about the problems associated with release of radioactive material into the environment and how that has a mutagenic impact on body cells. Yet we, as humans, have failed to see that the global pollution event that is plastic pollution may also have far reaching impacts beyond those that are obvious and observable to the naked eye. Maybe *this* is the problem – what we cannot see, we do not understand.

The World Health Organization's report draws on a number of case studies that examine biofilms – the thin layer of organic matter that forms on the surface of plastic – in microplastics residing in drinking water. We cannot usually see these biofilms with the naked eye.

They are, however, in almost every piece of pipework in our homes; they form in drink bottles, creeks, rivers and essentially anywhere that is a source of moisture and sustenance. Biofilms are usually inert and pose no threat to human health. The bacteria that live in biofilms are often beneficial and take on roles such as oxygenation of creeks and rivers. However, it has also been well documented that biofilms – including those on microplastic surfaces – can comprise a range of bacteria, pathogens and protozoa well known for making us sick, and the World Health Organization's report highlights a number of articles in the research literature that capture this issue. The biofilms are essentially a hotspot of bacteria, pathogens and protozoa that cause skin conditions, infections and even tissue and organ damage. While some of these free-living micro-organisms and pathogens may not present much concern to an otherwise healthy individual that has access to medical aid, many of the resultant illnesses associated with these micro-organisms and pathogens have the potential to be devastating in regions where poor diet and living conditions, and limited access to clean water and medical assistance is the reality. While we must of course highlight the importance of drinking water for a healthy diet, we must ensure that the water that is available through municipal-treated supplies and that which is available untreated through non-municipal supplies is free from these new sources of potential concern for disease transmission.

Plastic Pollution and Public Health

Due to the current limited, but emerging, knowledge about the behaviour of bacteria and pathogens, we must begin to question the role that plastics play as vectors of disease in some of those highly plastic-polluted regions that are also facing some significant health crises. Could micro- (or macro-) plastics that have found their way to rivers be responsible for the transport of deleterious large-scale diseases like Ebola in Africa? Could the seasonal outbreaks of plague and severe dysentery in Madagascar be intensified by the transmission of pathogens through the environment while rafting on plastic fragments? Some evidence exists to suggest that plastic pollution could be contributing to similar global health crises, and a number of research teams have called for a greater body of research in order to understand the problem. They have all settled on the conclusion that microplastics are probably acting as vectors of disease.[7] A study that considered nurdles (small plastic balls that are the raw material used to make many plastic products) at public swimming beaches found that they act as little rafts for *E. coli* and *Vibrio* species.[8] While these bacteria can occur naturally in the ocean, they are associated with diarrhoea and food poisoning in humans. These nurdles are unlikely to be ingested by humans at these beaches, thereby creating a direct pathway of exposure to these pathogenic bacteria, but they show that there is the possibility for pathogens to raft through the oceanic and aquatic environment on plastics. This is important when dealing with disease mitigation in those regions, like Madagascar, where plastic pollution and pathogenic bacteria are abundant. What it also tells us is that it would not be impossible for plastics to act as rafts for other more serious diseases. Ebola was mentioned earlier as a potential concern – could plastic be aiding in the distribution of the disease? A study conducted in 2015, just as global attention was turning to plastic pollution, examined exactly that question. The study considered the ability for the Ebola virus, the virus responsible for the death of over 11,000 people in West Africa, to be translocated on plastic used in the medical response effort.[9] The study noted that the virus can stay active on surfaces in the tropical environmental conditions of West Africa for around four to five days, and more when the conditions of humidity are optimal. This is not to criticise the medical response or condemn the use of plastic in the effort to stop the spread of Ebola. Plastic is a vital part of that effort and we must not risk direct transmission and infection because of a fear of plastic. The point here is to show that if plastic is used in the medical response (i.e., when the virus is already rampant), then plastic too can act as a mode of transmission of the virus in its early onset. The study shows that, under the right conditions, the Ebola virus can persist on plastics in the environment and remain viable for a period of time sufficient to rapidly transport the virus around the environment and between unsuspecting humans. It is not a far stretch to imagine an individual drinking from a plastic bottle, unaware that they are infected by the disease, then discarding that bottle haphazardly into a river where it is transported many kilometres over a three to four day period before being collected and used by the next individual.

Plastics and COVID-19

When I first started to write this book in a small café in Wales in 2018, the world had no idea what it was about to face less than two years later. Sitting nestled in the corner of that small café, I was surrounded by people talking, eating, drinking, and enjoying each other's company. I was a traveller from the opposite side of the world and yet I never thought about the health status of the other people in that room, nor was I concerned that they may be carrying a virus unbeknown to them or others around that may have catastrophic impacts on the livelihoods and lives of the entire room. I never thought about the journey I had taken to arrive in that small café or the places, people and diseases that I may have encountered along the way. Communal disease transfer was never part of our psyche and so we were never worried about it. In 2020, in a matter of months the world was gripped by the outbreak of the COVID-19 pandemic and everything changed. The epidemic changed the way we operate in communal life around the world. No longer can we while away the day in a small café minding our own business, watching the world go by. We are now subject to virus screening, venue check-ins, time limits and limitations on the number of people that can be in a common space at one time. The world revolutionised overnight to become a frantic cleaning machine, phobic of spreading a disease that we knew precious little about.

As the COVID-19 pandemic continued, our knowledge of the causative agents responsible for the initial outbreak and subsequent spread around the globe increased. We were suddenly made aware of the highly scientific-sounding name SARS-CoV-2, the virus responsible for the COVID-19 disease, and suddenly we had a wave of armchair microbiologists and virologists. People that had no formal training or pre-existing knowledge became experts in virus transmission, detection of the virus in the environment, and preventing community infections – or so they thought. We also saw a large number of experienced and well-informed scientific organisations begin to explore the way that SARS-CoV-2 could be distributed between individuals, through the environment and on surfaces. This latter transmission vector – transmission on surfaces – was, and still is at the time of writing, hotly refuted by health organisations and governments throughout the world. In contrast to the dogged position held by many decision makers, the scientific evidence for the transmission of SARS-CoV-2 on hard surfaces, of which plastic is included, has increased.[10] The strong evidence, which can be seen in the peer-reviewed literature, is that plastic is responsible for harbouring and possibly ultimately rafting COVID-19 around the world.[11] There were a number of examples of community transmission of unknown origin that could be explained in no other way than by plastic packaging acting as vector for virus transmission. It is now understood that the virus can remain viable for tens of days if environmental and surface conditions are optimal. A research study undertaken by the Australian Commonwealth Science Industry Research Organisation (CSIRO) showed that in controlled laboratory conditions, the SARS-CoV-2 virus remained viable for many weeks.[12] This finding was in contrast to the New South Wales Health regulator's stance at the time that the SARS-CoV-2 virus did not remain viable on hard surfaces.

We have learned through the COVID-19 pandemic that plastics do indeed act as vectors for the most severe and deadly viruses and diseases. A hypothetical world where diseases could be exacerbated by the rafting of bacteria, viruses and pathogens between countries has now become a reality.

CHAPTER TWELVE

Some Plastics We Can Live Without

I n June 2018 I was asked to contribute a piece to the popular academic media platform *The Conversation*.[1] I had previously provided comment to the Australian media that there are some single-use plastic products that we need and cannot live without. My argument was framed around the idea that in our current world, some plastics are essential, such as those used in medicine (e.g., syringes), those used in a limited capacity for water bottles in areas where drinking local water is unsafe, and even some food packaging circumstances where items are required to be preserved and stored for long periods such as in disaster aid situations. This was the first time that many in the media had thought about the necessity of some single-use plastics. The article explored the situations where we are left with no other option but to rely on single-use plastics. Many other media outlets wanted to run this idea as a story because it provided balance and a reality-check to the debate about single-use plastics that had not previously been considered. There is without doubt a need for certain single-use plastics in our current world.

Single-Use Plastics in Medicine

Many commentators on *The Conversation* article disagreed with the argument that in some situations and circumstances we cannot, and should not, avoid single-use plastics. Some suggested a return to glass and steel devices for medical applications. Thinking specifically of syringes, returning to reusable glass and steel syringes would create an unreasonable and unacceptable infection control risk, it would limit their ability to be sterilised on demand, and it would severely limit their ability to be easily transported to where they are needed.

In October 2021, there was a renewed outbreak of the Ebola virus disease in the Democratic Republic of Congo.[2] Single-use plastic medical devices provide the infection control capacity to fight this disease through vaccine administration and medical treatment. Imagine a remote field hospital that is already struggling to manage a patient load and provide effective treatment options, and then imagine that field hospital also having to sterilise glass syringes and other medical devices that are currently made from single-use plastics. In this instance, the need for fast and efficient medical devices required to preserve human life far outweighs the need to reduce the use of single-use plastics.

Managing Waste from Essential Plastics

Earlier in this book we examined the role that plastics play as vectors and rafts of bacteria and pathogens through the environment. It is all about a balance of risks. In order to reduce the spread of certain disease pathogens and virus, single-use plastic is the only sensible option. But the sensible option has a very non-sensible burden – the waste stream. So while it is important for single-use plastics to remain in service until another suitable alternative can be found, we must also manage that waste stream correctly. As an example, if life-saving medical waste was to be discarded carelessly after use, then it would create a long-term environmental burden that cannot justify the continued reliance of the health care industry on single-use plastics. The medical waste must be managed so that it is processed in a way that allows for the plastics to be recycled without danger of spreading disease, or destroyed in a system that ensures that it cannot perpetuate disease but also prevents the emission of dangerous by-products into the environment.

One approach (albeit not the most effective) to this could be the incineration of medical waste using a range of highly technical air filters called scrubbers to capture any of the gaseous phase pollutants. The point here really is that in the current global market, we have a product that works and does so well but it has environmental limitations and adverse implications; we have a need to shift away from the existing technology, but this can only really occur once there is a suitable substitute in place; and finally there is a need for more research – and funding to drive the research – and large-scale shifts in mindset required to transform the existing industry status quo.

A Greener Future

Fortunately we are already seeing some movement towards reducing plastic and other waste in medical care. A number of companies have formed in recent years that focus on creating reusable pouches, bags and medical equipment that can be easily sterilised in field conditions.[3] Many companies now also exist that manufacture medical devices from eco-friendly, plant-based and non-virgin feedstock materials. In doing so, they are creating products that may be able to be recycled, composted or have a low resource demand during manufacturing. One international organisation, Health Care Without Harm, has a strategic mission to provide healthcare globally that is environmentally and socially focussed.[4] A major focus of the organisation's work is to partner with companies that are able to provide environmentally-friendly medical equipment. It is organisations like these that help us to reduce our plastic emission burden, even when that plastic is essential.

The Plasticology Project   161

Mindful Use of Essential Plastics

In the responses to the media coverage of *The Conversation*, one of the commentators was so displeased with my proposition that in some circumstances we require single-use plastics that they called into question a link between myself and single-use plastic manufacturers, particularly the bottled water industry. I have never received any funding from such industry groups. The article and the sentiments were entirely a statement on reality and the fact that in the world, on some occasions we do need single-use plastics because they perform so well at the task they were designed to do: be single use.

But just because we need these single-use items, it doesn't mean that we can use them recklessly and without consideration of how we manage their end-of-life stages. Indeed, those single-use plastic items that we really need should be treated with extreme caution and the highest level of respect. In order to allow these single-use plastics to continue to be used to their full utility, it is important that they do not create a waste stream that is either irresponsible or negligent. How do we do this in a world where single-use plastics are discarded haphazardly into the natural world, often with very little consideration for their final destination? How do we expect a remote field hospital in the Democratic Republic of Congo to be able to manage copious volumes of single-use plastics generated during the process of fighting disease? This comes down to a balance. The balance must involve using existing single-use plastic technology until new and improved technology is available. Perhaps that technology can be plastic that is truly environmentally degradable. I use the term 'environmentally degradable' here because this is what plastic needs to become. The plastic must remain shelf-stable in order for it to remain useful as a product in medical applications or emergency food provision, but it must also be able to degrade into inert materials once it is placed into the environment, whether that is haphazardly through reckless disposal or through the intentional deposition into landfills and other repositories. This is where single-use plastic technology needs to be. Currently, some single-use plastics are marketed as biodegradable or compostable. The reality is that many of these products do not ever fully break down in the environment into inert materials. They instead break down into smaller components of themselves, or have no change even after extended periods of time in the environment.

One of the biggest challenges about attempting to reach the stage where the product is environmentally degradable is to avoid the temptation to use plastics that are derived from sources that could be otherwise used as a food supply or could become another environmental problem. It is important to avoid plastics that are made from materials that could result in the need for crop monocultures, such as corn for cornstarch, similar to the way ethanol for fuel has seen increased demand for certain crops. To achieve a plastic product that is both single-use and environmentally degradable will require a multi-faceted approach to polymer development and most likely a great deal of co-operation between stakeholders and funding organisations.

A Single Action to Reduce Plastic Pollution

A shift away from single-use plastic drinking bottles is possibly one of the best efforts that we can make to reduce the introduction of plastic pollution into the global environment. In September 2019, two journalists from *The Guardian* produced a visual representation of the impact that single-use plastic drinking bottles are having on our world.[5] This ground-breaking and shocking piece of journalism perfectly described the global thirst for single-use plastic drinking bottles through a series of cleverly constructed graphical representations. The journalists drew on meaningful and relevant references that allow us to fully comprehend the enormity of the problem. For example, the journalists tell us that every hour, 54.9 million bottles are purchased globally which equates to a pile as high as Rio de Janeiro's Christ the Redeemer statue. They also tell us that each day, 1.3 billion single-use plastic bottles are sold, and that if they were all piled up, the pile would be equivalent to the size of the Eiffel Tower. Staggeringly, they also describe one full years' worth of bottles as being taller than the Burj Khalifa in Dubai (the world's tallest building in 2021), with 481.6 billion bottles sold (based on 2018 figures). If you have ever seen this building in real life, then you will understand how tall this is – I remember standing at the base of the building, craning my neck towards the sky and still not being able to see the top. It is a remarkable structure. In a 10-year period, the journalists report that over 4 trillion single-use plastic bottled were sold, equating to a pile that engulfs Long Island City and Greenpoint, shadowing the skyscrapers of lower Manhattan Island.

As with other facets of plastic pollution, single-use plastic drinking bottles are a multi-component problem. The issue stems from production, but also from the demand of consumers. It has only been in recent years that consumers have grappled with the possibility that there may be alternatives to virgin plastics for use in plastic bottles. The journalists note that between 1950 and 2015, over 4.9 billion plastic bottles have been discarded and without proper data relating to their whereabouts, they have likely either ended up in landfill or discarded care-lessly into the global environment.

Some Plastics Are Essential, Others Are Not

Some plastics that we use in everyday life are, for now, essential. In the current world, despite all the best efforts of science, research and product development, we are not yet in a position where we can put an end to plastic manufacturing for good. For that reason, we must consider how we can best manage the plastics that we need, their end-of-life waste phases, and ensure that we continue to look to the future with the brightest of minds on the case to develop new technologies and solutions, so that we can eventually phase out the use of the majority of plastics in our lives.

CHAPTER THIRTEEN

The Solution to Pollution

Dilution or Prevention?

I used to work for a boss who would always tell me that 'the solution to pollution is dilution'. Perhaps so, if the pollution is in a concentration low enough to allow for dilution, but when the pollution is so large that it becomes a conspicuous part of your existence, what do you do? Well, in that case, maybe the solution to pollution isn't dilution, but rather *prevention*. As we have seen, plastic pollution has taken over the world. There is no place on Earth (or even near-Earth for that matter) that escapes the problems of plastic pollution. We know that plastic doesn't break down at any rate sufficient for it to be called degradable, and we know that almost every piece of plastic we create and use today will remain in the natural environment for generations after we are long forgotten.

So what is the solution? Prevention. But what does prevention mean in terms of a pollution source that is truly ubiquitous? Prevention takes many forms, but in recent times, the most widely considered way to reduce plastic pollution is to prevent the primary introduction of plastic products. This means a shift away from single-use plastics including shopping bags, packaging and straws. We now understand that if we remove the input of redundant plastic into the world, we also reduce the amount of output plastic that we must deal with. Remember, though, plastic itself is not inherently bad; it is the way in which we use and dispose of the plastic that makes it troublesome to our natural world. So there must be a balance, and as we saw in the last chapter, there are some applications of single-use plastics, such as in medicine, we currently really cannot go without.

Global Awareness of Plastic Pollution

As we, a global community, become more aware of the role we play in the production and reduction of plastic pollution, conversations on the topic are increasing and we are becoming far more aware of our own plastic use and that of others around us. Remarkably in recent years, using single-use plastics has become somewhat of a global criminal offence. I can recall a time when single-use plastic was used for everything and it was handed out like it was an infinite resource that had no ramifications for the health of the global natural environment. I once worked in a supermarket where we were not just encouraged but *required* by management to provide a single-use plastic bag to each customer regardless of their purchase size.

I would often be asked by customers for more bags, to double bag the heavy items or for some they could take home to line their rubbish bins. I remember visiting fast food outlets and cafés and each and every drink was adorned with a single-use plastic straw. But now, times and attitudes have changed. Scarcely anywhere in the world will offer a single-use plastic bag, let alone allow for double bagging and a take-home stash. Fast food outlets and cafés are banning plastic straws in an effort to reduce the overall generation of plastic waste in our daily lives. Many people are doing this on an individual level, too. It only takes a short search on social media platforms like Facebook to uncover a host of pages and groups that support a plastic-free and zero-waste lifestyle.

But scan through the comments and feedback on social media platforms, and very soon you realise that the plastic-free and zero-waste world is pretty brutal and judgemental. Despite their best efforts, those who want to replace their old cling-film for in-vogue beeswax wraps are often made to feel inferior to those who have already made the shift to plastic-free food coverings, or an entirely plastic-free life. There has emerged a culture of shaming those who still use a plastic-handled toothbrush and who don't shop at bulk food stores with reusable jars. But while some may sit and cast judgement on those who aren't yet as savvy with their plastic-free lifestyle, we must ask ourselves a question: can we really make a global difference just by using one less piece of plastic to cover our lunch?

Reducing Plastic Is a Challenge

The world of plastic-free living and reducing waste isn't always easy, but this may be a good thing. While it may feel like a frivolous task, these small changes are an indication of a bigger global shift. No longer is it culturally ok to use plastic in the same wasteful way as we have done for almost half a century. Plastic, particularly single-use plastics used for everyday goods, have become a global evil and this can be seen in the passion of those who use social media platforms to encourage others to do better. As a scientist I have even been the target of this. I often provide comment to the media on various issues around plastic pollution and a piece published by the *Sydney Morning Herald* in Australia regarding plastic pollution in drinking water attracted this comment: '... Dr Harvey apparently hasn't caught up with the scale of the problem [of plastic pollution].' I found this attack rather amusing and ironic given that I was drafting the manuscript for this book at the time!

There is certainly a code of conduct that needs to be adhered to as we cannot expect a message to be communicated and received when it is dripping with negativity. We need to be having conversations rooted in compassion and the desire to educate. We see this approach reflected in the multiple global campaigns to reduce plastic consumption, from National Geographic's Planet or Plastic? campaign, to the screens of BBC with its documentary *Drowning in Plastic*. These media pieces introduce us to the concept of plastic pollution, walk us through the evidence of the problem, show us the impacts of the problem and then educate us to do better and improve the situation for the natural environment. We, as a global civilisation, have now identified that plastic pollution is a problem and we need to do something about it. We understand that we all need to make changes, from our daily decision to package our lunch in a box rather than a plastic bag, or our decision to use a refillable water bottle rather than single-use plastic bottles. We know that the changes we make today to *reduce* our plastic emissions will ultimately mean that there is a reduction in the amount of plastic entering into the global environment.

Legislated Action to Reduce Plastic Pollution

As mere consumers, we aren't alone in our quest to reduce our day-to-day consumption of plastic. Despite some initial push back, many governments around the world are beginning to investigate single-use plastic and plastic pollution-related bans. Let's not fool ourselves into thinking that governments are doing this out of the goodness of their hearts. Well, not generally speaking, anyway. Plastic production and manufacturing is still a derivative of one of the most powerful and politically influential industries in the world – the oil and gas industry – and there is plenty to be lost by taking such bold steps away from plastics. This was perfectly highlighted in Australia when industry representatives and mining companies, among others, were asked by the government to provide opinions about the decline in the retail sector. A Department of the Treasury report that consolidated these stakeholder comments and was distributed to government ministers in April 2019 cited bans on single-use plastic bags as being a hindrance to shoppers, making them less able to transport larger quantities of groceries and therefore causing them to purchase fewer items at the grocery store.[2] With some of the largest donors to Australian political parties being companies either directly involved in the oil and gas industry or standing to gain from industries that benefit from the prolonged use of single-use plastic bags in retail settings, there is little wonder how the Department of the Treasury reached such a peculiar hypothesis about the decline in the performance of the retail sector.

Despite this blatant attempt to make the public rethink and become critical of the single-use plastic bag ban, the ban continues. At the end of 2021 all states and territories in Australia, with the exception of New South Wales, had implemented a single-use plastic bag ban. New South Wales expects to phase out single-use plastics bags by June 2022. Notably, there have been no reported cases of starvation resultant of shoppers being prevented from obtaining a single-use plastic bag. After all, there are always reusable shopping bags.

The influence of the oil and gas industry in the perpetua-
tion of plastic use does not stop at plastic bags. In a 2018
report by the International Energy Agency titled 'The
Future of Petrochemicals', it was modelled that plastics
will be the largest growth industry for oil and gas refining
products up to the year 2050. A letter from the American
Chemistry Council, an industry group representing the
oil, gas and manufacturing industries, to the Office of
the United States Trade Representative provided further
transparency to the intentions to grow the plastics industry
in many parts of the world, using Kenya as the gateway
through a new trade agreement:

*'Chemicals comprise about 17 percent of all goods
exports to Kenya. More than 80 percent of chemicals
exported to Kenya are resins: polyvinyl carbonate and
high-density polyethylene (HDPE). ... We anticipate
that Kenya could serve in the future as a hub for
supplying U.S.-made chemicals and plastics to other
markets in Africa through this trade agreement. Kenya
is growing its ground transportation network of rail-
ways and roads, has a port that meets international
standards in Mombasa on the Indian Ocean, and
boasts three international airports. Kenya's increasing
logistics capacity can support an expansion of chemicals trade **not just between the United States
and Kenya, but throughout East Africa and the
Continent.**[3] [emphasis added]*

The voices of industry representatives are often very loud on the topic of plastic pollution. In most instances, the interest that particular industry representative groups hold in ensuring the narrative around plastic is both positive and complementary of the material and market is evident. It is very interesting to reflect on a clash that occurred between Dr Rebecca Altman and the Plastics Industry Association (PIA) in early 2022. The PIA is a powerful industry body that represents plastic manufacturers and associated industries in the USA and throughout the world. On January 4, 2022, *The Atlantic* published Dr Altman's piece on plastics, 'How Bad Are Plastics, Really?'[4] The article standfirst read: 'They're harmful to health, environment, and human rights – and now poised to dominate this century as an unchecked cause of climate change.'

In a reply to this, the PIA ran a piece on their website, 'Oversimplified Arguments and Cherry-Picked Data Won't Solve Plastic Waste', with the standfirst: 'A recent article from *The Atlantic* uses selective evidence to criticise plastics without weighing the material's many benefits – from advancing sustainability goals to improving economic circularity and more.'[5]

This response was a blow-by-blow countering of the details put forward in Dr Altman's *Atlantic* article. The hypercritical response to Dr Altman, while uncomfortable to read, is a clear example of the polarising nature of the conversation about plastic and its place in the world.

A Changing Market

Fortunately, environmentally-aware governments have started to realise the importance of protecting the global environment. Global plastic pollution has become so synonymous with the current global civilisation that nobody wants to be remembered as the one that did nothing. Whatever their motive, dozens of countries around the world have now introduced some form of legislation or ruling to restrict the use and distribution of single-use plastic products. There is a very good chance that in whatever country you are reading this book, there will be a single-use plastic ban there. Much of this legislative banning pertains to the use and distribution of things like single-use plastic bags, single-use packaging, plastic straws, and single-use cutlery. Almost every day as I read the news, I find yet another story of a government restricting the use of these items. This is fantastic to see and it demonstrates the power of a mass cultural movement to solve a problem. We are now seeing supermarkets that previously did not buy into the plastic pollution problem changing their practices to provide consumers with reusable bags or bags made from recyclable materials. Those same supermarkets that previously felt it necessary to wrap every piece of fruit in a single-use plastic package are now leaving the fruit unwrapped on the shelf, wrapped in its own packaging just as nature intended it to be presented.

China has been used throughout this book as an example of a country that has the capacity to change the entire course of global plastic pollution simply by curbing its own use of single-use plastics. The economy of scale in China means that almost any effort to legislate change in the way that plastic is used and distributed in the country would almost instantly create a noticeable change in plastic waste production. Interestingly, one of the best solutions to plastic pollution that I have uncovered was actually in China. In a small, hole-in-the-wall style shop in a small town in a central province, I ordered a frozen dessert of hawthorn berries served on a skewer. When the dessert was handed to me, it had this paper-like film wrapped around the outside. Curiously, I gazed at the film, wondering if I was to peel it away or eat it. After a few moments of solitary contemplation, I was informed that this was in fact a type of edible paper! It was made from the remnants of wheat processing and flattened out into a tasteless, edible paper. These wrappers allowed the product to be protected, freezer stable and sold as fast food, all while being plastic free. Imagine that! If the global market harnessed this very simple technology, then there would be no telling how soon it would help curb the global plastic pollution crisis!

CHAPTER FOURTEEN

The Corporate Drive to Reduce Plastic Pollution

In recent times we have also seen huge global corporations like McDonald's shift away from single-use plastics in their restaurants. In Australia, McDonald's announced that it would move away from plastic straws by 2020, however this received some backlash by consumers.[1] So why did this particular decision enrage consumers so much? Well, as it turns out, most consumers really didn't like the idea of sipping their thickshake through a paper straw. I guess for those people, it will be a bit of tough luck.

Some of the more serious responses had to do with consumers who *need* to use a plastic straw, such as people with disabilities. Single-use paper straws tend to fall apart quite quickly and are easy to bite through, so pose a safety risk for people with limited jaw control. Reusable straws need to be washed well and regularly, which may be hard for some, and metal and silicone straws are often rigid, making drinking difficult or unsafe for people with mobility challenges. We must ensure that we have a scheme in place that allows those who legitimately need these products to be able to access them. Rapid knee-jerk reactions by governments to completely phase out all single-use plastics, without proper consideration of those legitimate users, will in time generate even more challenges for those users. However, I applaud McDonald's for taking such a strong step in the right direction. It is easy for a big global corporation to back away from such initiatives. For a corporation that is so ingrained in the global cultural mindset through marketing and advertising, it would be easy for them to instil in their customers the need to use a plastic straw. Instead, McDonald's listened to what the wider global consumer climate was telling them, and made a change.

Supply and Demand

Simple economics tells you that if you don't have a demand, then there really isn't much use for a supply. Ultimately, the outcome of banning these products would be to remove them from the marketplace altogether. This would remove the temptation to return to these products once the next global cultural movement comes along.

But despite this global momentum, there are some lawmakers around the world who are averse to the idea of placing a ban on single-use plastic products, including single-use plastic bags and straws. In the USA, the state of Arizona has legislated against banning single-use plastic items.[2] Similarly, Alabama has proposed a ban on the banning of single-use plastic products.[3] It is difficult to draw a conclusion about why these two states are making a deliberate effort to counter the global trend in restricting the supply of single-use plastic products, however, a look at these states' economies may hold the answer. Plastic is primarily derived from hydrocarbons: oil. In both Alabama and Arizona, oil takes a pivotal role in the economy: Alabama has a well-established oil industry, while Arizona has a number of potential oil reserves. This may or may not be the reason behind the ban on plastic bans, but there is certainly a link between the industry and some of those people who have attempted to propagate the laws that perpetuate the generation of single-use plastics and ultimately plastic pollution. By placing a ban on plastic bans, a sales avenue for oil-derived products is maintained.

Clean-ups

But why not just clean up the pollution that we already have and be more responsible with the new plastic waste that we are creating? Well, it really isn't that simple. The question once again returns to how do we collect and recover all that waste, and then what do we do with that plastic once we've recovered it? There isn't a simple answer to that. A group of 30 companies from around the world – including Dow, ExxonMobil, Shell, PepsiCo and many others, uniting as the Alliance to End Plastic Waste – recently pledged US$1 billion to help curb plastic pollution. Through investment in clean-up programs, research, and technologies to assist in recycling of plastic waste, the companies hoped that this contribution would act in a positive way to curb global plastic pollution. Of course, this announcement received some criticism by market commentators, who argued that many of these companies are responsible for the production of the plastic products in the first place. Perhaps they are, but we must also be careful not to jump too quickly on the bandwagon of corporation bashing. As with governments that ban an array of single-use plastic products, the motives of some of the contributing corporations may not always be necessarily pure, but that shouldn't detract from the opportunities for further progress that these gestures afford the global environment. We must look at each opportunity as a chance to take a step, big or small, in the right direction. We cannot all be working on the prevention end of the spectrum, sometimes it is still necessary to work on the clean-up and recovery end, but so long as we are all working towards the same common goal, we will eventually achieve it.

Every Little Bit Matters

We are starting to build a picture here: a picture of plastic pollution as a multi-faceted problem and one that there is no single solution for. But we can each do our little bit to help solve the problem. I am reminded of a quote from primatologist and anthropologist Dr Jane Goodall: 'Every individual matters. Every individual has a role to play. Every individual makes a difference.'

Whether talking about habitat protection for chimpanzees in the African jungle or cleaning up your local beach, the message still remains the same. We cannot continue to live in a world where we perceive our own actions to be of limited causality. We, as a civilisation, have a role to play in making sure that we protect the planet and provide a home for future generations.

In many ways, I look at the current plastic pollution movements – grass roots uprisings combating plastic and environmental pollution – around the world and smile. I smile because it restores my hope in the current dominant civilisation of Earth. It restores my hope in thinking that we *do* understand the sentiment that Dr Goodall so eloquently put forward; we *do* understand that we need to make changes in our own lives and in our world to make sure that we leave this planet in a cleaner state than we found it. Each time I see a photo of a team of young people spending their weekend at the local beach collecting plastic debris, I am blown away by just how much the global plastic pollution crisis has captured the hearts of the current generation. We may not be able to change the majority mindset of older generations, but our best hope is to change the mindset of the younger generations. This has been demonstrated repeatedly through history. Younger generations will be the big changers in our world; they will be the big thinkers, the problem solvers.

Already, we are seeing people from younger generations developing ideas to solve the plastic pollution problem. The Great Pacific Garbage Patch is receiving a clean-up by a device invented by Boyan Slat, a young man in his late teens at the time of invention. The device, while initially requiring some engineering tweaks, is responsible for skimming one of the most plastic-polluted parts of the ocean of its pollution load. This single invention has the potential to remove thousands of tonnes of waste from the global oceans, subsequently helping reduce the amount of suspended plastic matter in the ocean. Removing the suspended plastic matter means that fewer animals will mistake the plastic pollution for food, which means that less plastic is introduced into the global food chain. This also means that there are fewer opportunities for plastic to be transported around the world, and ultimately, we come one step closer to solving the problem of global plastic pollution. By the thoughtful invention of a device by an *individual*, the entire world is benefiting.

Computers as a Means of Reducing Plastic Pollution

We live in a very unique time in history where computer-based technology has become the central axis of our lives. In this computer age we have seen the global economy revolutionised. Trade is no longer done using banknotes and coins; electronic currency, or cryptocurrency, has changed the way global economics operates. Single units of cryptocurrency can now represent tens of thousands of units of the strongest tangible global currency. This has been revolutionary in the fight against plastic pollution.

On Manila Bay beach in the Philippines, often referred to as the 'toilet bowl', cryptocurrency was used to pay locals a wage per unit weight of waste that was collected from this beach.[4] In just a few months, the beach went from a dumpsite of plastic debris, mattresses, baby nappies and all sorts of other waste, back into a beach. Using cryptocurrency to pay for the beach clean-up was powerful because many of these local people were not wealthy and they often didn't have access to any formal banking system. By using a cryptocurrency system, the small NGO Bounties Network was able to pay workers while also demonstrating to doners where donations and funding were being used. This effectively allowed donor funding to be directly transferred to those locals on the ground who were doing the work.

This cryptocurrency scheme allowed the local people to supplement their income through cleaning the beach area, but also learn the *value* in the materials that can be recycled and reused. The ultimate aim of the exercise was to improve the condition of the environment the Philippines and surrounding oceans which are a major source of food and livelihood for the Filipino people.

CHAPTER FIFTEEN

The Plasticology Project

'The idea of the wilderness needs no defense, it only needs more defenders. Remaining silent about the destruction of nature is an endorsement of that destruction.'

These are the words of Edward Abbey from his 1977 book *The Journey Home,* a collection of essays describing the American West and Abbey's journey from urban landscapes into the 'natural world'. Abbey was 62 at the time of his death in 1989 and is often described posthumously as an anarchist, environmentalist and key contributor to the contemporary environmental protection movement. Abbey was a believer in giving the environment a voice to fight the battles it could not fight itself.

As you reflect on the content of this book and consider how *The Plasticology Project* may be applicable to your own life, it may be difficult and overwhelming to think about how you can take action to become a defender of the environment and natural world, and not an endorser of its destruction. Fortunately, when it comes to dealing with the challenges of plastic pollution, there is a huge number of individuals, community groups, not-for-profits, and corporations that have already started to fly the environmental flag with the intention of being environmental defenders. This chapter will hopefully inspire you to take up the call to care for our planet – our only home – and join others in the fight for protection of the environment and the natural world, and to become a Plasticology Project Ambassador in your home, your community or your country.

We don't all have to make grand gestures and become the next Greta Thunberg standing on the global activist stage; we can act locally and still have a global impact. If we all do *what* we can *where* we can, then together we can slowly begin to improve the global environment, the natural world and reverse the plastic crisis that we are facing.

Each of us can make the exact same difference in the world. Perhaps we may not invent some large apparatus that can be towed through the oceans, but we can each *individually* do our bit. You can pick up a piece of plastic or three, you can start a clean-up campaign, or you can post online. Every little bit matters; every *individual* matters.

Think About It This Way

If you see a chocolate bar wrapper in the roadside gutter, pick it up and put it in a rubbish bin. By doing that one action you have prevented that wrapper from being washed down the drain, into the stormwater network, out into a river, and into the ocean where it would be transported across to the other side of the world, broken down into microplastics and enter the food chain. Your one simple action that takes only a few seconds has the potential to prevent an entire cascade of events. In light of this, let's now take a look at some of the local- and global-scale actions that you could be a part of as a person committed to being a defender of the environment through *The Plasticology Project*. This is not meant to be an exhaustive list, but rather a few examples to inspire you to get involved and make changes in your own life.

Plastic-Free Lifestyle

As we have already seen in this book, being part of the plastic solution does not have to be a big, grand statement on the global scale. Each of us can make a difference simply by making a few changes in our personal and homes lives, no matter where you live.

- Instead of using a plastic single-use shopping bag, opt instead for a jute, hemp or other natural fibre bag that can be reused and eventually, when no longer functional, returned to the Earth via composting.

- Carry a refillable water bottle for use throughout the day; let's not forget about all of those free water fountains that we mentioned in Chapter 7!

- Reusable containers, beeswax wraps, metal or paper drinking straws and refillable coffee cups are all the norm in plastic-conscious homes now, and they provide great alternatives for reducing waste, reducing plastic use and shrinking our plastic footprint.

One great challenge that my family and I often set ourselves at home is to go for a week without using *any* plastic packaging. At first this is a real challenge, but as the week progresses, your eye sharpens and suddenly you are aware of all the products that do contain plastic packaging and, conversely, those that do not. Need a new toothbrush? You could use the same plastic handled, plastic bristled, plastic packaged toothbrush that you've always used, or you could try something new – a bamboo toothbrush perhaps? While the bristles tend to be nylon (plastic), the handle and package are almost certainly made from compostable bamboo and recyclable card-board. Need mushrooms? Forget about the pre-packaged mushrooms on the shelf that are covered in plastic, fill up a paper bag or your own homemade cotton veggie bags. This simple switch means no plastic and probably a cheaper price too.

The great thing about challenging ourselves to a plastic-free life is that there are so many alternatives out there on the market, so many people that have the same passion to reduce their plastic and environmental footprint, so many people that are already a part of *The Plasticology Project*, that it is really very easy once you get started. If you need some extra inspiration on how to reduce your plastic footprint in your life and in your home then it is worthwhile taking a look at the work of some influencers in this space, including Bea Johnson, author of *Zero Waste Home*; Lauren Singer, founder of Trash is for Tossers; and Anne-Marie Bonneau, founder of Zero Waste Chef.

Join a Community Group or Campaign

So now you have started on the journey of becoming an ambassador of *The Plasticology Project* and you're searching for a way to make a difference in your local area. This is great news! Just like participating in the movement at the household level, there are many ways that you can get involved at the community level. A quick search on the internet or social media is often a great way to find like-minded individuals and community groups. If you can't find a community to engage with, then maybe this is your opportunity to start your own. In Australia, we have some exceptionally driven individuals who have created a community of like-minded environment defenders and plastic pollution warriors.

Boomerang Alliance is one of the largest organisations in Australia with a focus on reducing the human environmental burden through multiple avenues.[1] In 2017, they commenced the Plastic Free Noosa project, and this involved the engagement of multiple stakeholders including councils, local tourism organisations, media, community groups and many others. The purpose of the project was to reduce plastic use within the Noosa community in Queensland, educate retailers about plastic-free options, encourage non-landfill approaches to waste management and promote concepts of a circular economy. The program signed on 204 businesses, and in 2019 had eliminated 4.3 million pieces of plastic from use, which would likely have gone to landfill or into the environment. The program was so successful that the Plastic Free Places organisation was launched and as of 2022 there are 12 projects in operation throughout Australia. As each project trial comes to an end, the projects are handed over to the local community for continuation into the future.

Take 3 for the Sea was conceived by marine ecologist Roberta Dixon-Valk and youth educator Amanda Marechal in 2009. Joined by environmentalist Tim Silverwood in 2010, Take 3 for the Sea quickly became a social movement in Australia. Driven by a desire to reduce the amount of marine plastic pollution, the trio set out to educate the community about plastic pollution and to promote the simple action of people taking three pieces of rubbish with them as they leave the beach. The idea being that if each person collects three pieces of rubbish, then as a collective, we will soon have the beaches free from plastic rubbish. Through a network of volunteers and outreach activities the Take 3 movement has become a global phenomenon with somewhat of a cult following.

Another example of an activist-led organisation that is making waves in Sydney with its efforts to reduce plastic pollution is **Ocean Tidings**, guided by environmental educator Sarah-Jo Lobwein. Based in the Sutherland region of Sydney, Australia, which is famous for Cronulla Beach, Ocean Tidings has a number of initiatives that strive to remove plastic pollution from the ocean. Through targeted social media campaigns like #swapforthesea and #rejectasingle (use plastic), Ocean Tidings has built a network of dedicated community followers who actively participate in rubbish clean-ups of the beaches, hold educational sessions and promote plastic-free retail and hospitality within the region.

I mention these three examples here because they are local to where I live in Sydney, Australia. All over the world there is an abundance of groups, activities and initiatives that all share the same common goal – to be defenders of the environment. If you are in New York, USA, you may consider becoming involved with an initiative like **Environment New York** which actively campaigns against plastic packaging in retail outlets and provides a 'plastic scorecard' for retailers.[2]

If you are in London, UK, you might investigate the #OneLess campaign created by **Marine CoLABoration** in 2016 with the aim of eliminating single-use plastic water bottles from London. As part of the initiative, 28 drinking water fountains have been installed across London to promote the use of re-fillable bottles and the consumption of fresh tap water. The campaign encourages participants to pledge through social media that they will stop using single-use plastic water bottles, with these being replaced by refillable bottles and drinking tap water. The initiative is estimated to have removed 5 million single-use plastic bottles and 9 million single-use plastic items from retailers and suppliers since its inception in 2016.[3]

If you live in Kenya and are searching for a community to become involved with, then you might be interested in joining the community created by James Wakibia (we met James earlier in Chapter 5). In 2015, James set out to raise awareness about plastic pollution near his home town of Nakuru by launching a Twitter campaign with the hashtag #banplasticsKE. James has been recognised by the United Nations Environment Programme for his effort to reduce plastic pollution in Kenya and his instrumental role in improving the environment in Africa overall.[4] In 2017, the Kenyan government responded to James by banning single-use plastic bags in the country. Among many other initiatives, James has now created an online community of over 21,000 Twitter followers who all receive regular updates about the status of plastic pollution in Kenya. James' work is a great demonstration of how an individual can upscale efforts from one single person with a passion to an entire community movement.

Start an Initiative

You may think that starting your own initiative as an environment defender and ambassador of *The Plasticology Project* is a bit overwhelming and not for you, but if you do decide that you would like to take on the challenge it can be incredibly rewarding. The most important thing to remember when starting is to have a passion for the cause. If you have read this far in the book then there is a good chance that you have that passion already!

Inspiration for starting your own initiative can come in many forms. In 2020, pre-eminent artist Dr Yu Fang Chi presented a sculpture piece at a FORM art exhibition in Perth, Australia. The piece, titled *Remnant III*, is an intricately-formed overhead installation of plastic shopping bags that serve to represent fish swimming through the ocean.[5] The installation is a reminder that we are 'drowning in a tide of plastic'.[6] As the bags gently sway in the breeze, the imagination can easily see how the scene depicts the analogy. Remarkably, Dr Chi did not use plastic shopping bags for this installation, but instead remnant fabric from a textile factory was cut and sewn to resemble plastic bags. Dr Chi's work casts light on many important global issues related to plastic pollution, overuse of resources, environmental stewardship and responsible product lifecycles. Through the work, Chi introduced these concepts to a wider audience, created a community around these concepts, and contributed in a unique way to the global discussion on plastic pollution. Dr Chi's work demonstrates that being an environment defender and an ambassador of *The Plasticology Project* can take many different personalised forms.

Working through an existing community, online or social media presence is another great way to start a new initiative and promote its goals. This is exactly what Dan and Kika from **Sailing Uma** have done. Through their already strong social media presence that showcases their journeys sailing around the globe in their 'eco-yacht', Dan and Kika have also managed to highlight the issue of plastic pollution in the oceans by engaging and teaming up with the environment-defending organisation, **In the Same Boat**. Via a series of special short feature films,[7] Dan and Kika have been able to show the work that is being done by In the Same Boat to recover plastic pollution from remote parts of the northern oceans. In the Same Boat have an ambitious plan to clean 20,000 Norwegian beaches through their work.[8] In addition to beach cleans, they are a source of knowledge and educational resources that allow their community to connect, raise awareness and help in the battle against plastic pollution. This is a great example of how communities can build on each other for a common cause, just like being an ambassador in *The Plasticology Project*.

Another great grass-roots initiative that caught my attention while researching this book is the **Plastic Free India** campaign led by India Youth for Society.[9] It is estimated that in India only around 40% of the plastic waste generated each day (approximately 26,000 tonnes) is collected and managed in a responsible manner, with the remaining being discarded into the environment, rivers and oceans.

The Plastic Free India campaign has three primary aims: to raise awareness about plastic pollution in India, to provide affordable alternatives to plastics, and to establish a recycling facility to sort and manage waste collected from the target regions. One of the key features of the campaign is generating employment for local community members through waste collection and recycling – often called 'rag pickers'. This is traditionally an informal employment sector with workers being paid piecemeal rates for their work. This campaign strives to commoditise responsible management of plastic pollution and promote principles of a circular waste economy.

Global Plastic Pollution Reduction Strategies

The Plasticology Project is all about taking action where we can, building a community and supporting each other as ambassadors and defenders of the environment. Each step and each action that we take adds to the global movement that is laser focussed on reducing global plastic pollution. Whether we are working as an individual, as part of a local community or as part of a larger initiative, we are contributing to a narrative that is driving change at the global scale. We can see this in the emergence of global, multi-country, multi-organisation initiatives that are tasked with ambitious targets for combating plastic pollution.

Circular Economy

The idea of circular economy has become popular among globally-influencing organisations, corporations and governments. The idea of a circular economy is not new, however its application to combating the plastic pollution crisis is relatively fresh. In a perfect circular economy world, all of the plastic materials that have been produced to date would be able to be returned to a raw material state and then used once again to generate a new product. Organisations acting on the global scale like the **Ellen MacArthur Foundation** are champions of a plastic circular economy.[10] The foundation has three simple guiding principles:

1. Eliminate all problematic and unnecessary plastic items.

2. Innovate to ensure that the plastics we do need are reusable, recyclable or compostable.

3. Circulate all the plastic items we use to keep them in the economy and out of the environment.

The foundation has established the **Plastics Pact Network**, a global network of organisations that have subscribed to reducing their plastic footprint. The Australia, New Zealand and Pacific Islands (ANZPAC) arm of the Plastics Pact Network has 60 founding members including major corporations Nestlé, Coca-Cola South Pacific, PepsiCo, and Unilever.

Through the network, the foundation is calling for a binding United Nations treaty for a circular economy to eliminate plastic pollution.[11]

Policy Change

Affecting policy change is one of the best ways to become an environment defender and ambassador for *The*

Plasticology Project. It is actually much easier than you may think to have an influence on policy and legislation that govern manufacturing, materials, waste, and plastic pollution reduction. Each of the examples provided in this chapter can be directly linked to policy change. Whether you are making choices in the home to purchase plastic-free products, or you are a part of a community movement that calls for a ban on single-use plastic bags, you are sending a message that policies and regulations need to change now.

It really is as easy as making a change in your life. This same philosophy is adopted by the **#breakfreefromplastic** movement. Originating in the USA in 2016, the movement is now a global initiative that engages 11,000 organisations and individuals. The movement promotes different approaches for managing waste, encourages community engagement, provides a network of like-minded individuals, but most importantly aims to influence and inform policy change at local, country and international levels.[12]

While sharing similar sentiments as most other waste reduction movements, the #breakfreefromplastic movement is very clear in its stance that waste incineration is not a solution to managing plastic waste, which is currently a popular strategy in many parts of the world.[13]

What Next?

We have reached the end of this book, but not the end of the journey. I hope that you have found this book to be insightful and useful. As I ponder over the contents of this book, I feel a sense of weight, a burden to do more. We are in a unique time in history where we are presented with a number of challenges that could have very serious repercussions for the planet and ourselves. Often it feels overwhelming to be presented with the reality of the global plastic pollution crisis, but it is comforting to know that there is hope for change in the near future.

Unique to this period, we are aware of exactly what those challenges are and we have a reasonably good comprehension of how we might go about fixing them. We have seen in this book a number of examples of how passionate and dedicated individuals, communities and organisations are contributing in a positive way to the reduction of plastic pollution. But despite the good news stories that emerge from time to time to give our environmental conscience a pat on the back, we are still faced with a bleak future. The question is not *if* humans will desecrate the planet and the natural world in a way that is irreparable, but *when*.

As the species on the planet that holds itself in highest esteem, and by many accounts is the most intelligent species, we do very well at being ignorant of the warning signs of impending danger. Humans are being presented the clearest of warning signs that our planet is in peril, and yet we still go to great lengths to turn away and ignore those warnings. Some special humans have figured it out. Those who have been in a deep connection with the natural world for many decades can see and read the warning signs – others cannot. We, as a species, are inherently ignorant to the blatant truths that are right in front of our eyes. Globally we have seen the impacts of plastic pollution, we have seen the way that ecosystems have been desecrated by the presence of a material that is so pervasive and ubiquitous that it is, in one form or another, inside every living organism on the planet. Yet, it is only a select few who can truly *see* the impact that this pollution is having on our only home. Those select few drive local, national and global campaigns to raise awareness and educate others.

I hope that by reading this book you have been educated and enlightened. I hope that you have discovered something that triggers a change – a drive to do more. I hope that you feel compelled to stand up and take action and to do your bit to change the world. For if we continue on the trajectory that we have been following, then the outcome is clear. Our legacy to future civilisations will not be great temples and intricate artefacts, it will not be great works of philosophy or even remarkable artworks. The legacy of this civilisation for future civilisations will be plastic pollution. Whether found at the depths of the ocean, or the farthest reaches of the universe, the Anthropocene will be remembered as the Plasticene – the age of plastic. This is not the end of the journey – far from it. The journey is only just beginning. We are now all a part of *The Plasticology Project* and we have work to do! So, let's work together to make the small changes in our lives that become the catalysts for even bigger changes on the global scale. Together, as a global community of ambassadors, we can beat plastic pollution and leave behind a legacy of being the generation that helped solve the plastic pollution crisis. One small change at a time.

Notes

INTRODUCTION

1   Rachel Carson, *Silent Spring* (New York: Houghton Mifflin, 1962).

2   Linda J. Lear, 'Rachel Carson's "Silent Spring",' *Environmental History Review* 17, no. 2 (1993): 23–48, https://doi.org/10.2307/3984849.

3   'The Great Smog of 1952,' Met Office, accessed May 1, 2022, https://www.metoffice.gov.uk/weather/learn-about/weather/case-studies/great-smog.

4   Peter G. Wells, 'The Iconic *Torrey Canyon* Oil Spill of 1967 – Marking Its Legacy,' *Marine Pollution Bulletin* 115, no. 1–2 (February 2017): 1–2, https://doi.org/10.1016/j.marpolbul.2016.12.013.

5   Edward J. Carpenter and K. L. Smith, 'Plastics on the Sargasso Sea Surface,' *Science* 175, no. 4027 (March 1972): 1240–1241, https://doi.org/doi:10.1126/science.175.4027.1240.

6   'Blue Planet II,' BBC Earth, accessed May 1, 2022, https://www.bbcearth.com/shows/blue-planet-ii.

7   'Planet or Plastic?,' National Geographic, accessed May 1, 2022, https://www.nationalgeographic.com/environment/topic/planetorplastic.

1 PLASTIC

1   L. H. Baekeland, 'The Synthesis, Constitution, and Uses of Bakelite,' *Journal of Industrial & Engineering Chemistry* 1, no. 3 (March 1909): 149–161, https://doi.org/10.1021/ie50003a004.

2   Leo H. Baekeland. Method of Making Insoluble Products of Phenol and Formaldehyde. US Patent 942,699, filed July 13, 1907, and issued December 7, 1909.

3   Kai Zhang et al., 'Understanding Plastic Degradation and Microplastic Formation in the Environment: A Review,' *Environmental Pollution* 274 (April 2021): article 116554, https://doi.org/10.1016/j.envpol.2021.116554.

2 THE GLOBAL PLASTASTROPHE

1   Imogen E. Napper and Richard C. Thompson, 'Environmental Deterioration of Biodegradable, Oxo-Biodegradable, Compostable, and Conventional Plastic Carrier Bags in the Sea, Soil, and Open-Air Over a 3-Year Period,' *Environmental Science & Technology* 53, no. 9 (April 2019): 4775–4783, https://doi.org/10.1021/acs.est.8b06984.

2   Jenna R. Jambeck et al., 'Plastic Waste Inputs from Land into the Ocean,' *Science* 347, no. 6223 (February 2015): 768–771, https://doi.org/doi:10.1126/science.1260352.

3   Zongguo Wen et al., 'China's Plastic Import Ban Increases Prospects of Environmental Impact Mitigation of Plastic Waste Trade Flow Worldwide,' *Nature Communications* 12, (January 2021): article 425, https://doi.org/10.1038/s41467-020-20741-9.

3 THE LINKS

1   Alejandra Borunda, 'Ocean Warming, Explained,' *National Geographic*, August 14, 2019, https://www.nationalgeographic.com/environment/article/critical-issues-sea-temperature-rise.

2   Thomas R. Knutson et al., 'Tropical Cyclones and Climate Change,' *Nature Geoscience* 3 (February 2010): 157–163, https://doi.org/10.1038/ngeo779.

3   Thomas R. Knutson et al., 'Tropical Cyclones and Climate Change,' *Nature Geoscience* 3 (February 2010): 157–163, https://doi.org/10.1038/ngeo779.

4   Weather is used in this context to disambiguate climate change on the global scale and localised climatic variability.

5   Andrés Cózar et al., 'The Arctic Ocean as a Dead End for Floating Plastics in the North Atlantic Branch of the Thermohaline Circulation,' *Science Advances* 3, no. 4 (April 2017), https://doi.org/doi:10.1126/sciadv.1600582.

4 OCEAN PLASTICS ARE A GLOBAL CHALLENGE

1   Britta Denise Hardesty et al., 'Estimating Quantities and Sources of Marine Debris at a Continental Scale,' *Frontiers in Ecology and the Environment* 15, no. 1 (February 2017): 18–25, https://doi.org/10.1002/fee.1447.

2   Adam Morton, 'Seven Tonnes of Marine Plastic Pollution Collected on Remote Arnhem Land Beach,' *The Guardian*, September 6, 2019, https://www.theguardian.com/environment/2019/sep/06/seven-tonnes-of-marine-plastic-pollution-collected-on-remote-arnhem-land-beach.

3   L. Lebreton et al., 'Evidence That the Great Pacific Garbage Patch Is Rapidly Accumulating Plastic," *Scientific Reports* 8 (March 2018): article 4666, https://doi.org/10.1038/s41598-018-22939-w.

4   L. Lebreton et al., 'Evidence That the Great Pacific Garbage Patch Is Rapidly Accumulating Plastic," *Scientific Reports* 8 (March 2018): article 4666, https://doi.org/10.1038/s41598-018-22939-w.

5   John E. Dore et al., 'Summer Phytoplankton Blooms in the Oligotrophic North Pacific Subtropical Gyre: Historical Perspective and Recent Observations,' *Progress in Oceanography* 76, no. 1 (January 2008): 2–38, https://doi.org/10.1016/j.pocean.2007.10.002.

6   *Journey to Planet Earth*, directed and written by Hal Weiner (Washington DC: Screenscope Inc., 2009).

7   Sarah Gibbens, 'Plastic Proliferates at the Bottom of World's Deepest Ocean Trench,' *National Geographic*, May 14, 2019, https://www.nationalgeographic.com/science/article/plastic-bag-mariana-trench-pollution-science-spd.

8   Gabriel Erni-Cassola et al., 'Distribution of Plastic Polymer Types in the Marine Environment: A Meta-Analysis,' *Journal of Hazardous Materials* 369 (May 2019): 691–698, https://doi.org/10.1016/j.jhazmat.2019.02.067.

5 RIVERS – THE MISSING LINK

1   Cristiano Rezende Gerolin et al., 'Microplastics in Sediments from Amazon Rivers, Brazil,' *Science of The Total Environment* 749 (December 2020): article 141604, https://doi.org/10.1016/j.scitotenv.2020.141604.

2   Ling Yang et al., 'Microplastics in Freshwater Sediment: A Review on Methods, Occurrence, and Sources,' *Science of The Total Environment* 754 (February 2021): article 141948, https://doi.org/10.1016/j.scitotenv.2020.141948.

3   'Our Story,' River Cleanup, accessed May 1, 2022, https://www.river-cleanup.org/en/our-story.

4   Wenya Luo et al., 'Comparison of Microplastic Pollution in Different Water Bodies from Urban Creeks to Coastal Waters,' *Environmental Pollution* 246 (March 2019): 174–182, https://doi.org/10.1016/j.envpol.2018.11.081.

5   Laurent C. M. Lebreton et al., 'River Plastic Emissions to the World's Oceans,' *Nature Communications* 8 (June 2017): article 15611, https://doi.org/10.1038/ncomms15611.

6   Lin Zhu et al., 'Microplastic Pollution in North Yellow Sea, China: Observations on Occurrence, Distribution and Identification,' *Science of The Total Environment* 636 (September 2018): 20–29, https://doi.org/10.1016/j.scitotenv.2018.04.182; Xiaoxia Sun et al., 'Microplastics in Seawater and Zooplankton from the Yellow Sea,' *Environmental Pollution* 242, Part A (November 2018): 585–595, https://doi.org/10.1016/j.envpol.2018.07.014; Xiong Xiong et al., 'Microplastics in the Intestinal Tracts of East Asian Finless Porpoises (*Neophocaena Asiaeorientalis Sunameri*) from Yellow Sea and Bohai Sea of China,' *Marine Pollution Bulletin* 136 (November 2018): 55–60, https://doi.org/10.1016/j.marpolbul.2018.09.006.

7   Daniel González-Fernández et al., 'Floating Macrolitter Leaked from Europe into the Ocean,' *Nature Sustainability* 4 (June 2021): 474–483, https://doi.org/10.1038/s41893-021-00722-6.

8   Giuseppe Suaria et al., 'The Mediterranean Plastic Soup: Synthetic Polymers in Mediterranean Surface Waters,' *Scientific Reports* 6 (2016): article 37551, https://doi.org/10.1038/srep37551.

6 PLASTIC ANIMALS

1   'Africa Is on the Right Path to Eradicate Plastics,' UN Environment Programme, posted July 30, 2018, https://www.unep.org/news-and-stories/story/africa-right-path-eradicate-plastics.

2   'Woolworths Ooshies Warning: Steer Clear of People Selling Lion King Collectables Online,' 7 News, last modified August 1, 2019, https://7news.com.au/lifestyle/woolworths/woolworths-ooshies-warning-steer-clear-of-people-selling-lion-king-collectables-online-c-375537.

3   S. C. Gall and R. C. Thompson, 'The Impact of Debris on Marine Life,' *Marine Pollution Bulletin* 92, no. 1–2 (March 2015): 170-179, https://doi.org/10.1016/j.marpolbul.2014.12.041.

4   Jennifer L. Lavers and Alexander L. Bond, 'Exceptional and Rapid Accumulation of Anthropogenic Debris on One of the World's Most Remote and Pristine Islands," *PNAS* 114, no. 23 (May 2017): 6052–6055, https://doi.org/doi:10.1073/pnas.1619818114.

5   Jennifer L. Lavers, Ian Hutton, and Alexander L. Bond, 'Ingestion of Marine Debris by Wedge-Tailed Shearwaters (*Ardenna Pacifica*) on Lord Howe Island, Australia During 2005–2018,' *Marine Pollution Bulletin* 133 (August 2018): 616–621, https://doi.org/10.1016/j.marpolbul.2018.06.023.

6   Jennifer L. Lavers, Ian Hutton, and Alexander L. Bond, 'Ingestion of Marine Debris by Wedge-Tailed Shearwaters (*Ardenna Pacifica*) on Lord Howe Island, Australia During 2005–2018,' *Marine Pollution Bulletin* 133 (August 2018): 616–621, https://doi.org/10.1016/j.marpolbul.2018.06.023.

7   Krista M. Verlis, Marnie L. Campbell, and Scott P. Wilson, 'Seabirds and Plastics Don't Mix: Examining the Differences in Marine Plastic Ingestion in Wedge-Tailed Shearwater Chicks at Near-Shore and Offshore Locations,' *Marine Pollution Bulletin* 135 (October 2018): 852–861, https://doi.org/10.1016/j.marpolbul.2018.08.016.

8   Google Scholar analytics: search terms "seabird" AND "marine" AND "plastic".

9   J. F. Provencher et al., 'Garbage in Guano? Microplastic Debris Found in Faecal Precursors of Seabirds Known to Ingest Plastics,' *Science of The Total Environment* 644 (December 2018): 1477–4784, https://doi.org/10.1016/j.scitotenv.2018.07.101.

10 Christopher K. Pham et al., 'Plastic Ingestion in Oceanic-Stage Loggerhead Sea Turtles (*Caretta Caretta*) Off the North Atlantic Subtropical Gyre,' *Marine Pollution Bulletin* 121, no. 1–2 (August 2017): 222–229, https://doi.org/10.1016/j.marpolbul.2017.06.008.

11 Katharine E. Clukey et al., 'Persistent Organic Pollutants in Fat of Three Species of Pacific Pelagic Longline Caught Sea Turtles: Accumulation in Relation to Ingested Plastic Marine Debris,' *Science of The Total Environment* 610–611 (January 2018): 402–411, https://doi.org/10.1016/j.scitotenv.2017.07.242.

12 Jennifer L. Lavers et al., 'Entrapment in Plastic Debris Endangers Hermit Crabs,' *Journal of Hazardous Materials* 387 (April 2020): article 121703, https://doi.org/10.1016/j.jhazmat.2019.121703.

13   Peter G. Ryan et al., 'Rapid Increase in Asian Bottles in the South Atlantic Ocean Indicates Major Debris Inputs from Ships,' *PNAS* 116, no. 42 (September 2019): 20892–20897, https://doi.org/doi:10.1073/pnas.1909816116.

14   Albert A. Koelmans et al., 'All Is Not Lost: Deriving a Top-Down Mass Budget of Plastic at Sea,' *Environmental Research Letters* 12, no. 11 (November 2017): article 114028, https://doi.org/10.1088/1748-9326/aa9500.

15   Winnie Courtene-Jones et al., 'Consistent Microplastic Ingestion by Deep-Sea Invertebrates over the Last Four Decades (1976–2015), a Study from the North East Atlantic,' *Environmental Pollution* 244 (January 2019): 503–512, https://doi.org/10.1016/j.envpol.2018.10.090.

7 MINISCULE PLASTIC

1   Andy Kollmorgen, 'Tap Water Vs Bottled Water,' CHOICE, last modified August 22, 2016, https://www.choice.com.au/food-and-drink/drinks/water/articles/tap-water-vs-bottled-water.

2   *I Nasoni di Roma, e nel mondo!,* v. 2.7 (Fabrizio DiMauro, 2020). Available to download on the Apple App Store.

3   Kate Browne, 'Is Bottled Water Safer Than Tap?' CHOICE, last modified March 21, 2018, https://www.choice.com.au/food-and-drink/drinks/water/articles/is-bottled-water-safer-than-tap-water.

4   Eid I. Brima, 'Physicochemical Properties and the Concentration of Anions, Major and Trace Elements in Groundwater, Treated Drinking Water and Bottled Drinking Water in Najran Area, KSA,' *Applied Water Science* 7 (2017): 401–410, https://doi.org/10.1007/s13201-014-0255-x.

5   Sherri A. Mason, Victoria G. Welch, and Joseph Neratko, 'Synthetic Polymer Contamination in Bottled Water,' *Frontiers in Chemistry* 6 (September 2018), https://doi.org/10.3389/fchem.2018.00407.

6   Justine Barrett et al., 'Microplastic Pollution in Deep-Sea Sediments from the Great Australian Bight,' *Frontiers in Marine Science* 7 (October 2020), https://doi.org/10.3389/fmars.2020.576170.

7   Katsiaryna Pabortsava and Richard S. Lampitt, 'High Concentrations of Plastic Hidden Beneath the Surface of the Atlantic Ocean,' *Nature Communications* 11 (August 2020): article 4073, https://doi.org/10.1038/s41467-020-17932-9.

8   Ali Karami et al., 'The Presence of Microplastics in Commercial Salts from Different Countries,' *Scientific Reports* 7 (April 2017): article 46173, https://doi.org/10.1038/srep46173.

9   Ji-Su Kim et al., 'Global Pattern of Microplastics (MPs) in Commercial Food-Grade Salts: Sea Salt as an Indicator of Seawater MP Pollution,' *Environmental Science & Technology* 52, no. 21 (October 2018): 12819–12828, https://doi.org/10.1021/acs.est.8b04180.

10   Philipp Schwabl et al., 'Assessment of Microplastic Concentrations in Human Stool – Preliminary Results of a Prospective Study,' *United European Gastroenterology Journal* 6, no. S8 (October 2018): A127.

11   D. J. Perez-Venegas et al., 'First Detection of Plastic Microfibers in a Wild Population of South American Fur Seals (*Arctocephalus Australis*) in the Chilean Northern Patagonia,' *Marine Pollution Bulletin* 136 (November 2018): 50–54, https://doi.org/10.1016/j.marpolbul.2018.08.065.

12   Mary J. Donohue et al., 'Evaluating Exposure of Northern Fur Seals, *Callorhinus Ursinus*, to Microplastic Pollution through Fecal Analysis," *Marine Pollution Bulletin* 138 (January 2019): 213–221, https://doi.org/10.1016/j.marpolbul.2018.11.036.

13   Rana Al-Jaibachi, Ross N. Cuthbert, and Amanda Callaghan, 'Up and Away: Ontogenic Transference as a Pathway for Aerial Dispersal of Microplastics,' *Biology Letters*, 14, no. 9 (September 2018):article 20180479, https://doi.org/doi:10.1098/rsbl.2018.0479.

14   A. F. R. M. Ramsperger et al., 'Environmental Exposure Enhances the Internalization of Microplastic Particles into Cells,' *Science Advances* 6, no. 50 (December 2020), https://doi.org/doi:10.1126/sciadv.abd1211.

15   Heather A. Leslie et al., 'Discovery and Quantification of Plastic Particle Pollution in Human Blood,' *Environment International* 163 (May 2022): article 107199, https://doi.org/10.1016/j.envint.2022.107199.

16   Lauren C. Jenner et al., 'Detection of Microplastics in Human Lung Tissue Using µFTIR Spectroscopy,' *Science of The Total Environment* 831 (July 2022): article 154907, https://doi.org/10.1016/j.scitotenv.2022.154907.

8 PLASTIC POLLUTION AT EXTREMES

1   "The World of Air Transport in 2018," International Civil Aviation Organization, accessed May 1, 2022, https://www.icao.int/annual-report-2018/Pages/the-world-of-air-transport-in-2018.aspx.

2   Qantas, 'Qantas Operates World's First Zero Waste Flight,' news
     release, May 8, 2019, https://www.qantasnewsroom.com.au/
     media-releases/qantas-operates-worlds-first-zero-waste-flight.

3   'Zero Waste & Plastic-Free,' SFO, 2022, accessed May 1, 2022, https://
     www.flysfo.com/environment/zero-waste-plastic-free.

4   Alfredo Carpineti, 'Something Terrible Is Happening on Mount
     Everest – and It's Our Fault,' IFLScience, posted June 18, 2018, https://
     www.iflscience.com/environment/something-terrible-is-happening-on
     -mount-everest-and-its-our-fault.

5   Imogen E. Napper et al., 'Reaching New Heights in Plastic Pollution
     – Preliminary Findings of Microplastics on Mount Everest,' *One
     Earth* 3, no. 5 (November 2020): 621–630, https://doi.org/10.1016/j.
     oneear.2020.10.020.

9 SPACE

1   'Space Debris and Human Spacecraft,' NASA, last modified May
     28, 2021, https://www.nasa.gov/mission_pages/station/news/
     orbital_debris.html.

2   United Nations Office for Outer Space Affairs, *United Nations Treaties
     and Principles on Outer Space*, ST/SPACE/11 (New York: UN, 2002), 6,
     https://www.unoosa.org/pdf/publications/STSPACE11E.pdf.

3   Mark Garcia, 'Station Crew Preps for Space Debris Avoidance,'
     NASA, posted September 22, 2020, https://blogs.nasa.gov/spacesta-
     tion/2020/09/22/station-crew-preps-for-space-debris-avoidance
     -maneuver.

4   Jamie Smyth, 'US Start-up LeoLabs Maps Out Plan to Make Dollars
     from Space Junk,' *Financial Times*, September 30, 2018.

5   Justin St P. Walsh and Alice C. Gorman, 'A Method for Space
     Archaeology Research: The International Space Station Archaeological
     Project,' *Antiquity* 95, no. 383 (August 2021): 1331–1343, https://doi.
     org/10.15184/aqy.2021.114.

10 PLASTICS IN SOIL

1   Haihe Gao et al., 'Effects of Plastic Mulching and Plastic Residue
     on Agricultural Production: A Meta-Analysis,' *Science of The Total
     Environment* 651, no. 1 (February 2019): 484–492, https://doi.
     org/10.1016/j.scitotenv.2018.09.105.

2   'Plastic Film Covering 12% of China's Farmland Pollutes Soil,' *Bloomberg*, September 6, 2017, https://www.bloomberg.com/news/articles/2017-09-05/plastic-film-covering-12-of-china-s-farmland-contaminates-soil.

3   'China Drafts New Rules to Control Rural Plastic Pollution,' *Reuters*, December 6, 2019, https://www.reuters.com/article/us-china-environment-plastic/china-drafts-new-rules-to-control-rural-plastic-pollution-idUSKBN1YA06T.

4   Ee-Ling Ng et al., 'An Overview of Microplastic and Nanoplastic Pollution in Agroecosystems,' *Science of The Total Environment* 627 (June 2018): 1377–1388, https://doi.org/10.1016/j.scitotenv.2018.01.341; Standards Australia, *AS 4454–2012: Composts, Soil Conditioners and Mulches* (Sydney: SAI Global Ltd, 2012).

5   Ee-Ling Ng et al., 'An Overview of Microplastic and Nanoplastic Pollution in Agroecosystems,' *Science of The Total Environment* 627 (June 2018): 1377–1388, https://doi.org/https://doi.org/10.1016/j.scitotenv.2018.01.341.

11 PLASTIC VECTORS

1   World Health Organization, *Microplastics in Drinking-Water* (Geneva: World Health Organization, 2019).

2   Joleah B. Lamb et al., 'Plastic Waste Associated with Disease on Coral Reefs,' *Science* 359, no. 6374 (January 2018): 460–462, https://doi.org/doi:10.1126/science.aar3320.

3   ARC Centre of Excellence Coral Reef Studies, 'Plastics Linked to Disease in Coral,' news release, January 26, 2018, https://www.coralcoe.org.au/media-releases/plastics-linked-to-disease-in-coral.

4   Joleah B. Lamb et al., 'Plastic Waste Associated with Disease on Coral Reefs,' *Science* 359, no. 6374 (January 2018): 460–462, https://doi.org/doi:10.1126/science.aar3320.

5   Erik R. Zettler, Tracy J. Mincer, and Linda A. Amaral-Zettler, 'Life in the "Plastisphere": Microbial Communities on Plastic Marine Debris,' *Environmental Science & Technology* 47, no. 13 (June 2013): 7137–7146, https://doi.org/10.1021/es401288x.

6   A. Mark Osborn and Slobodanka Stojkovic, 'Marine Microbes in the Plastic Age,' *Microbiology Australia* 35, no. 4 (October 2014): 207–210, https://doi.org/10.1071/MA14066.

7   Maocai Shen et al., 'Micro(nano)plastics: Unignorable Vectors for Organisms,' *Marine Pollution Bulletin* 139 (February 2019): 328–331, https://doi.org/10.1016/j.marpolbul.2019.01.004; Anutthaman Parthasarathy et al., 'Is Plastic Pollution in Aquatic and Terrestrial Environments a Driver for the Transmission of Pathogens and the Evolution of Antibiotic Resistance?,' *Environmental Science & Technology* 53, no. 4 (January 2019): 1744–1745, https://doi.org/10.1021/acs.est.8b07287.

8   Alyssa Rodrigues et al., 'Colonisation of Plastic Pellets (Nurdles) by *E. Coli* at Public Bathing Beaches,' *Marine Pollution Bulletin* 139 (February 2019): 376–380, https://doi.org/10.1016/j.marpolbul.2019.01.011.

9   Robert Fischer et al., 'Ebola Virus Stability on Surfaces and in Fluids in Simulated Outbreak Environments,' *Emerging Infectious Diseases* 21, no. 7 (July 2015): 1243–1246, https://doi.org/10.3201/eid2107.150253.

10   Faezeh Seif et al., 'The SARS-CoV-2 (COVID-19) Pandemic in Hospital: An Insight into Environmental Surfaces Contamination, Disinfectants' Efficiency, and Estimation of Plastic Waste Production,' *Environmental Research* 202 (November 2021): article 111809, https://doi.org/10.1016/j.envres.2021.111809.

11   Leslie Dietz et al., 2019 Novel Coronavirus (COVID-19) Pandemic: Built Environment Considerations to Reduce Transmission,' *mSystems* 5, no. 2 (April 2020), https://doi.org/10.1128/mSystems.00245-20; Neeltje van Doremalen et al., 'Aerosol and Surface Stability of SARS-CoV-2 as Compared with SARS-CoV-1,' *New England Journal of Medicine* 382, no. 16 (April 2020): 1564–1567, https://doi.org/10.1056/NEJMc2004973; Denis E. Corpet, 'Why Does SARS-CoV-2 Survive Longer on Plastic than on Paper?,' *Medical Hypotheses* 146 (January 2021): article 110429, https://doi.org/10.1016/j.mehy.2020.110429.

12   'How Long the Virus Can Survive,' CSIRO, last modified October 12, 2020, https://www.csiro.au/en/research/health-medical/diseases/covid-19-research/how-long-the-virus-can-survive.

12 SOME PLASTICS WE CAN LIVE WITHOUT

1   Paul Harvey, 'There Are Some Single-Use Plastics We Truly Need. The Rest We Can Live Without,' *The Conversation*, June 29, 2018, https://theconversation.com/there-are-some-single-use-plastics-we-truly-need-the-rest-we-can-live-without-99077.

2   'Ebola: North Kivu, Democratic Republic of the Congo, October–
    December 2021,' World Health Organization, accessed May 3,
    2022, https://www.who.int/emergencies/situations/ebola-outbreak
    -north-kivu-october-2021.

3   One such product is the EnviroPouch Reusable Steam Sterilization
    Pouch, https://www.enviropouch.com.

4   'Leading the Global Movement for Environmentally Responsible
    Health Care,' Health Care Without Harm, accessed May 3, 2022,
    https://noharm.org.

5   Simon Starr and Marco Hernandez, 'Drowning in Plastic: Visualising
    the World's Addition to Plastic Bottles,' Reuters Graphics, last modified
    September 4, 2019, https://graphics.reuters.com/ENVIRONMENT-
    PLASTIC/0100B275155/index.html.

13 THE SOLUTION TO POLLUTION

1   Sarah Berry, 'We "Eat a Credit-Card Worth of Plasitc" a Week. Do
    Water Filters Help?,' Sydney Morning Herald, August 26, 2019, https://
    www.smh.com.au/lifestyle/health-and-wellness/we-eat-a-credit-card-
    worth-of-plastic-a-week-do-water-filters-help-20190826-p52kse.html

2   Michael Roddan, 'Economy Falls Through Our Shopping Bags,' The
    Australian, June 24, 2019, https://www.theaustralian.com.au/nation/
    economy-falls-through-our-shopping-bags/news-story/c9fcea40c72
    881fa950f722a6c6500d0.

3   Ed Brzytwa, 'ACC Public Comments on U.S. Negotiating Objectives
    for a Trade Agreement with Kenya,' American Chemistry Council,
    posted April 28, 2020, https://www.americanchemistry.com/
    better-policy-regulation/trade/resources/acc-public-comments-on-
    us-negotiating-objectives-for-a-trade-agreement-with-kenya.

4   Rebecca Altman, 'How Bad Are Plastics, Really?' The Atlantic, January
    4, 2022, https://www.theatlantic.com/science/archive/2022/01/
    plastic-history-climate-change/621033.

5   'Oversimplified Arguments and Cherry-Picked Data Won't Solve Plastic
    Waste,' This Is Plastics, accessed May 5, 2022, https://thisisplastics.
    com/environment/oversimplified-arguments-and-cherry-picked-data-
    wont-solve-plastic-waste.

14 THE CORPORATE DRIVE TO REDUCE

1   Samantha Bailey, "McDonald's to Phase Out Plastic Straws in Australia by 2020," *The Australian*, July 18, 2018. https://www.theaustralian.com.au/business/news/mcdonalds-to-phase-out-plastic-straws-in-australia-by-2020/news-story/93247db9ad3f7d96175 26b41cb55541a.

2   Stephanie Morse, 'Fry's Foods Plans to Phase out Plastic Bags,' *Cronkite News*, October 8, 2018, https://cronkitenews.azpbs.org/2018/10/08/kroger-frys-food-will-phase-out-plastic-bags-over-7-years.

3   Sophia Waterfield, 'Alabama Seanate Pushes Legislation to Stop Plastic Bag Bans,' *Newsweek*, October 4, 2019, https://www.newsweek.com/alabama-vote-prohibit-plastic-ban-1391753.

4   Justin Calderon, 'Poorer Communities in the Developing World Bear the Brunt of Plastic Pollution. Could a New Digital Payment System Spark a Clean-up Revolution?' BBC Future, posted June 13, 2019, https://www.bbc.com/future/article/20190613-a-simple-online-system-that-could-end-plastic-pollution.

15 THE PLASTICOLOGY PROJECT

1   'Plastic Free Places,' Boomerang Alliance, accessed May 5, 2022, https://www.plasticfreeplaces.org.

2   'Tell Whole Foods: Planet Over Plastic,' Environment New York, accessed May 5, 2022, https://environmentnewyork.org/feature/nye/tell-whole-foods-planet-over-plastic.

3   #OneLess, *A Practical Guide to Tackling Ocean Pollution at Source*, September 2021, https://www.onelessbottle.org/wp-content/uploads/sites/14/2021/08/OneLess-toolkit-final.pdf.

4   'Meet James Wakibia, the Campaigner Behind Kenya's Plastic Bag Ban,' UN Environment Programme, posted May 4, 2018, https://www.unep.org/news-and-stories/story/meet-james-wakibia-campaigner-behind-kenyas-plastic-bag-ban.

5   Remnant III,' Yu Fang Chi, accessed May 5, 2022, http://yufangchi.com/installation.php.

6   Heather Jerrems, 'Plasticology Exhibition – An Interview with Yu Fang,' *Perth Happenings*, March 19, 2020, https://perthhappenings.com.au/plasticology-exhibition-an-interview-with-yu-fang.

7   Sailing Uma, 'You WON'T BELIEVE What We Found Out Here – Sailing Uma [Step 264],' May 7, 2021, YouTube video, 26:51, https://www.youtube.com/watch?v=RIJ4P8BT-Mc.

8   'The Fight Against Plastic,' In the Same Boat, accessed May 5, 2022, https://www.inthesameboat.eco.

9   'India's Youth Take on Plastic Pollution,' UN Environment Programme, posted July 8, 2022, https://www.unep.org/news-and-stories/story/indias-youth-take-plastic-pollution.

10 'Designing Out Plastic Pollution,' Ellen MacArthur Foundation, accessed May 5, 2022, https://ellenmacarthurfoundation.org/topics/plastics/overview.

11 'A UN Treaty on Plastic Pollution,' Ellen MacArthur Foundation, accessed May 5, 2022, https://ellenmacarthurfoundation.org/towards-a-un-treaty-on-plastic-pollution.

12 'Pushing for Policy,' Break Free From Plastic, accessed May 5, 2022, https://www.breakfreefromplastic.org/campaigns/pushingforpolicy.

13 'Who We Are,' Break Free From Plastic, accessed May 5, 2022, https://www.breakfreefromplastic.org/about.

Acknowledgements

The catalyst for writing this book came when I travelled to Wales in 2018. When I started putting my thoughts on a page back then, while sitting in a small café, I never thought that I would one day be writing the acknowledgment section for what is soon to become the published version of this book. That was a tumultuous year and writing about plastic pollution became my escape. What a journey it has been since those first words hit the page. The world has changed immensely since that day, but one thing remains the same and is clearer now more than ever – humans *need* to do better at protecting the natural world.

But this entire journey would not have happened if not for one person, Phoebe. I can confidently say that the pages in this book would have never made it out of the jumble of Word documents and notebooks if not for you. I consider myself so incredibly lucky to be able to share life and so many adventures with my best friend, my partner, and my rock. Writing this book has been one of those great adventures of life and you have stood by my side the entire journey. Phoebs, you have been an incredible light, life coach, spiritual guide, sounding board and grounder during the entire process. You kept me going during the highest highs and the lowest lows, and you kept me focussed on the light even during the dark times. You read early drafts, gave me feedback and listened to my incessant talking about 'the book', so thank you for going through it all. I know it wasn't always easy. I am grateful beyond words for your love and support that never fades, no matter the season of life. If not for your unconditional love through all of the journeys of our life, I may not ever have had the opportunity to write these words, so thank you. I am so glad that while we were walking along the beach in Aber, you said in your matter-of-fact way that hits like a freight train, 'Why don't you write a book about plastic pollution?' What a great idea, why didn't I think of that? I often say that many of *my* ideas come from our discussions about the world, life and everything in between. So, it must be said that you inspired this book, and so much of what surrounds it. For that I will be forever grateful. So, Phoebs, thank you for everything, always. I am truly lucky to have you; I love you.

I want to thank all the people that have been involved in the production of this book, Dixie Carlton and Ann Dettori Wilson of Indie Experts have been absolutely incredible taking the draft manuscript from concept through to published product. My sincere thanks to you both for seeing the potential in my words and crafting then into a *real* book. My thanks also to the wider production team at Indie Experts; you have all done such an amazing job to get the book to where it is today.

Thank you to my wonderful beta readers, Lynette Hodgson, Sue Martin, Professor Gabriel Filippelli and Sarah-Jo Lobwein. Your thoughts and comments were valuable in shaping the direction of this book.

Thank you to all the people – too many to name – that I have had conversations with over the years about the environment, the ocean, pollution and everything else that we need to start taking just a bit more seriously. Many of those conversations have sparked the thoughts in this book. It is so inspiring to know so many people are out there working towards the common goal of a better environment and future.

Thank you to my parents, Wendy and Peter. From a young age you opened the eyes of Nathan and I to a world different to our life in the suburbs. We would go on adventures to the beach, the mountains or the bush. A weekend drive was always in search of somewhere different to experience and see. You shaped my love for travel, of the outdoors and being lost in nature.

Finally, thanks to you, the reader. I am so honoured to think that you have taken the time to pick up my book and read what I have to say. I hope that you find something useful in this book and that you are inspired to make a change for the better in this world. We can all make a change, big or small, so please, enjoy the book and think about how we can all work together to create a better future for generations to come.

For More Information

If you would like to learn more about *The Plasticology Project*, and how you can help reduce the impact of plastic waste at individual, local, and global levels, please visit www.docpjharvey.com/theplasticologyproject.

For checklists and ideas for ways you can easily reduce your plastic consumption, please visit www.docpjharvey.com/ideas.

About Dr Paul Harvey

Following a childhood blessed with an abundance of outdoors exploration and with a strong desire to do more than 'just sit in a classroom', Dr Paul Harvey immersed himself in the study of environmental science from a young age.

He was propelled further along that path after a close family member was diagnosed with a rare form of cancer – one that was environmentally stimulated. Undertaking a PhD was the next step, and so the stage was set for a lifetime of wanting to make a difference.

Deeply passionate about the problems that plastics pose for our world, inspiration to write *The Plasticology Project* and share this passion with the world came after a day of picking up plastic waste on a beach in Wales with his partner.

Paul is driven by the desire to help others understand the issue of plastic, and encourage them to make mindful, positive changes for future generations.

Made in United States
Orlando, FL
18 July 2023

35266298R00130